2nd International Conference on Mechanical, Aeronautical and Automotive Engineering (ICMAA 2018)

MATEC Web of Conferences Volume 166 (2018)

Singapore
24 – 26 February 2018

Editors:

H.F. Abdul Amir
S. Kok Swee

ISBN: 978-1-5108-6198-5

Printed from e-media with permission by:

Curran Associates, Inc.
57 Morehouse Lane
Red Hook, NY 12571

Some format issues inherent in the e-media version may also appear in this print version.

Printed by Curran Associates, Inc. (2018)

For additional information, please contact EDP Sciences – Web of Conferences at the address below.

EDP Sciences – Web of Conferences
17, Avenue du Hoggar
Parc d'Activité de Courtabœuf
BP 112
F-91944 Les Ulis Cedex A
France

Phone: +33 (0) 1 69 18 75 75
Fax: +33 (0) 1 69 28 84 91

contact-edps@webofconferences.org

Additional copies of this publication are available from:

Curran Associates, Inc.
57 Morehouse Lane
Red Hook, NY 12571 USA
Phone: 845-758-0400
Fax: 845-758-2633
Email: curran@proceedings.com
Web: www.proceedings.com

TABLE OF CONTENTS

PREFACE

Unlike most conferences with a focused research interest and unlike the Mechanical Engineering Society meetings that target certain area, the second International Conference on Mechanical, Aeronautical and Automotive Engineering (ICMAA 2018) brings together researcher researchers in Mechanical Engineering from all over the world to share our results and achievements.

At first, we did not expect anything special with this arrangement, but as the meeting unfolds, we have noticed that different area or region of the world have very different interests. In particular, we have researchers from China and India. From their presentation, we see some very advanced researches that are in tune with the industry development in their region. It is both wonderful and awesome to see how the large economies have gradually move to connect their industry with academy and how the technology advancement in these countries are taking advantage of the research results. There are also research results on either very advanced applications of artificial intelligence on the mechanical systems. We observe how the smaller economies would try to leap forward with more advanced or supposedly next generation industry interests.

From the practical point of view, there is certain risk in this undertaking, but it does give the smaller economies chances to leap forward and take lead in certain area of technologies. I have observed many researchers from different area of interests got together and discussed possible fusion of their. I have also seen invitations for visits from and to various part of the world. This occurred not from the focused research interest; instead, it was from the fusion of interests that brought forth the kind of collaboration.

I am sure the International Conference on Mechanical, Aeronautical and Automotive Engineering will continue to play the row of providing researchers from the world the opportunity to expose their results and to know result from other areas.

Conference Chair

Prof. Jia-Yush Yen
National Taiwan University, Taiwan

Statement of Peer review

In submitting conference proceedings to MATEC Web of Conferences, I certify to the Publisher that I adhere to the **Policy on Publishing Integrity** of the journal in order to safeguard good scientific practice in publishing.

1. All articles have been subjected to peer review administered by the proceedings editors.
2. Reviews have been conducted by expert referees, who have been requested to provide unbiased and constructive comments aimed, whenever possible, at improving the work.
3. Proceedings editors have taken all reasonable steps to ensure the quality of the materials they publish and their decision to accept or reject a paper for publication has been based only on the merits of the work and the relevance to the journal.

Title, date and place of the conference

ICMAA 2018, The 2nd International Conference on Mechanical, Aeronautical and Automotive Engineering. Singapore, February 24-26, 2018.

Proceedings editor(s):

Dr. Haider Fa. Abdul Amir
Professor, Faculty of Applied Sciences
Universiti Teknologi MARA, Malaysia
haideral_amir@yahoo.com
haider@salam.uitm.edu.my

Prof. Sim Kok Swee,
The Multimedia University, Malaysia
kssim@mmu.edu.my

Date and editor's signature April 11, 2018

MATEC Web of Conferences **166**, 01001 (2018) https://doi.org/10.1051/matecconf/201816601001
ICMAA 2018

Studies on Mechanical Behaviour of Aluminium-Silicon Carbide-Copper Metal Matrix Composite

Shreyas P S[1], Shriramkumar S[2], Sruthisagar R[3], Vijaya Ramnath B[4] and Elanchezhian C[5]

[1, 2, 3] *Department of Mechanical Engineering, Sri Sairam engineering College, Chennai*
[4] *Professor, Dept. of Mechanical Engineering, Sri Sairam Engineering. College, Tamil Nadu, India*
[5] *Professor, Dept. of Mechanical Engineering, Sri Sairam Engineering. College, Tamil Nadu,India*
E- mail[1]:tryshreyas@gmail.com E-mail[4]: vijayaramnathb@gmail.com

Abstract. Metal matrix composites (MMC) are the combination of two or more dissimilar base metals which in turn gives new metal. Composites have better properties than their base material. Nowadays, composites play the major role in manufacturing industries especially in automotive industry. This paper reveals the reinforcement of Aluminium alloy (6061) with Silicon Carbide and Copper in different ratios to produce new composite using stir casting technique in normal atmospheric temperature. In this work, mechanical behavior of the composite namely tensile strength, hardness and Flexural strength were evaluated. It was observed that sample 2 shows highest tensile strength of 155 MPa with hardness 73. The result shows that the hybrid composite had better mechanical property.

1 Introduction

Nowadays composite materials are replacing conventional materials due to their high strength to weight ratio, bio degradable nature, wear and corrosion resistance. Mohan Kumar et al [1] found the Ultimate tensile strength of Al-4.5 wt% Cu-2 wt % Fly ash -3 wt%. SiC composites to be higher than the Base Matrix composite of Al -4.5wt%of Copper alloy. They stated that this material exhibits high strength to weight ratio and good fatigue resistance. Anilkumar et al [2] stated that the bonding between the matrix material and Fly ash is good and he also observed that there is no gap between the Particle and matrix Material. Itskos et al [3] discovered that by adding the Fly Ash to the MMC, it enhances the tribo-performances of the composites. Shivaprakash et al [4] found that the heat treated Fly ash composite has comparatively better characteristics to non heat treated composites, especially the wear performance is better for water quenched specimen than air cooled specimen and non-heat treated specimens. Arun Kumar et al [5] stated that the alumina coated stirrer blades will prevent the migration of ferrous ions from stirrer to the molten metal. They also found that the e-glass and Fly ash composites with aluminium (Al 6061) increase the tensile strength by 60-70%. Vivekananthan et al [6] stated that by adding the Fly ash and Mg in aluminum results in increased hardness from 58BHN to 86 BHN. Mahendra et al [7] found that the fluidity and density of the composites decreases while tensile strength, impact strength, compression strength and dry sliding wear resistantance increase due to increase in percentage of Flyash. Surendra Kumar Patel et al [8] discovered that both the reinforcements 50% SiC and 50% of Zircon sand on aluminium ADC-12 alloy give better compressive strength compared to others. Ramanpreet Singh and Rahul Singla [9] stated that by increasing the weight fraction of Silicon Carbide, the wear rate gets reduced linearly. Ashok Atulker et al [10] found that the wear rate remains constant more or less in the pressure range of 0.5N/mm^2 to 0.6N/mm^2, but the wear rate increases steadily up to the normal pressure of 0.5N/mm^2, and increases rapidly that due to adhesive wear. Vinod Kumar et al [11] stated that the wear rate of the composition (Al6061T6/15%SiC/15% Al2O3 MMC) is highly influenced by the applied load. Gaurav Mahajan et al [12] concluded that by adding SiC and TiB2 to the matrix material, the wear rate and coefficient of friction gets decreased. They also found that by adding SiC and TiB2 to the aluminium alloy (Al6061), Hardness is increased considerably. By using this Composition (AL/10%SiC) the hardness value increased by 38%, also by using the composition (Al/10%SiC/5%TiB2) the hardness value increases by 35.7. %. Francis Xavier et al [13] stated that, by increasing the Wet grinder Stone Dust particle (WSD) content from 10% wt to 20% wt in a composite, the hardness and wear resistance increases but the ductility reduces. Aluminium with silicon carbide as reinforcement has been done by various authors [14, 15]. Meena et al [16] stated that by increasing the reinforcement (SiC) size (220 mesh, 300 mesh and 400 mesh) the impact strength decreases. Omkar Pansare et al stated that the wettability property increases by increasing the Mg % in the melt. They also stated that by

https://doi.org/10.1051/matecconf/201816601001

preheating the mould, we can avoid porosity in casting and increase the fluidity and proper fluid distribution. Vijaya Ramnath et al [18,19, 20] studied the compression and chemical Properties of Aluminium Nano Composite and also reviewed aluminium metal matrix composites. Radhika et al [21] stated that the incorporation of Graphite as primary reinforcement, which forms the protective layer between the pin and the counter face, increases wear resistance. Manoj Singla et al [22] found that a sample which contains 25% SiC gives the best value in Hardness and Toughness, i.e., Hardness 45.5 BHN and Impact Strength 36 N-m. Beyond this weight fraction, the hardness decreases because SiC particles interact with each other resulting in cluster formation Since, no work has been carried out using Silicon Carbide-Copper as reinforcement for Aluminium, this paper aims to fabricate metal matrix composites using combination.

2 Materials used

Aluminium Alloy (6061): This alloy is manufactured based on the ASTM E 1251- 07 optical emission spectrometry. Density of this material is $2.65 g/cm^3$. This alloy has excellent corrosion resistance in both ordinary and marine conditions. By Anodic Treatment this property can be further enhanced. Al 6061 rods are used in this process that are melted before the reinforcement added to the molten metal.

Silicon Carbide: Silicon Carbide is the natural resource which melts at around 2,730 $^{\circ}$C and having hardness of 9.0-9.5 Mohs. Silicon Carbide powder is used in this Composite as a reinforcement.

Copper: Copper powder is used as reinforcement in this Composite. Copper usually melts at a temperature of 1500 $^{\circ}$C. Copper is known to possess good ductility, thermal conductivity and electrical conductivity.

3 Fabrication of composite

Required amount of raw Aluminum ingots (6061) will be taken in graphite crucible and it will be placed for 2 hours inside the furnace in 850-900^{0}c. At the same time the Silicon Carbide and Copper powder will get preheated in the same temperature. When the aluminum ingots melts completely, the Degasser added to the molten metal which removes all the gasses and impurities presents in the molten metal. Then the cover-all powder is added, which forms a layer on top of the molten metal and it retains heat inside the crucible. Then the preheated Silicon Carbide and Copper will be added in the molten metal. Now, the stirring process will be carried out for 10 minutes and the stirring temperature is maintained in 650-700^{0}C. To avoid casting defects the die will be preheated to 500^{0}C, when it reaches 500^{0}C the molten metal poured in to the die (100x100x10mm) and which is allowed to solidify. Table 1 shows composition of MMC fabricated in this work.

Table 1. Material Composition

Name of Sample	Percentage of Al 6061	Percentage of SiC	Percentage of Cu
Sample 1	94.25	0.75	5
Sample 2	93.5	1.5	5
Sample 3	92.5	2.5	5

4 Testing of composite

4.1 Tensile test: Tensile test of the fabricated hybrid composite sample was prepared as per ASTM: D638. This test was carried out in Universal Testing Machine, Which has done in the normal atmospheric temperature (35^{0}C) and controlled relative humidity.

4.2 Hardness test: Hardness is resistance to indentation and by which the hardness was determined by permanent depth of indentation. The hardness test of a material was done on Brinell Hardness Testing Machine, Avery make 3000 kgf and it was followed by ASTM E10 standard. This test was performed by 10mm ball by applying 3000 kgf load.

$$\text{Brinell Hardness (HB)} = \frac{2F}{\pi D(D-\sqrt{D-d})}$$

Where,

D = Ball diameter
d = Impression
diameter F = Load (N)
HB = Brinell hardness number

4.3 Flexural Test: The Flexural (Compression) test was done on Universal Testing Machine UTN 40 Model. Here, the test specimen was prepared as per the ASTM E8 Standard.

5 Results and discussions

5.1 Tensile Properties

The result of tensile test is shown in graph and Fig. 1, from which it is concluded that the sample-2 (1.5% Sic) has high strength than the other two compositions. This is due to the presence of Silicon Carbide, but not in enough of a quantity to skew the strength too much towards either side.

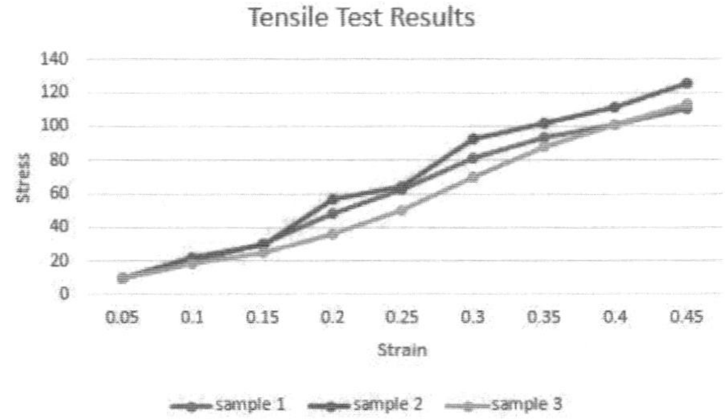

1(A) Results of Tensile Test- Graph

1(B) Result of Tensile Test- Specimen

5.2 Hardness Properties

Fig. 2 (A & B) show the result of hardness test, from which it is concluded that sample 2 which has 1.5% of Silicon Carbide and 5% of copper has more hardness than the other 2 samples. This is because Silicon carbide in higher quantities can make the composite brittle and here it is being balanced by the copper present.

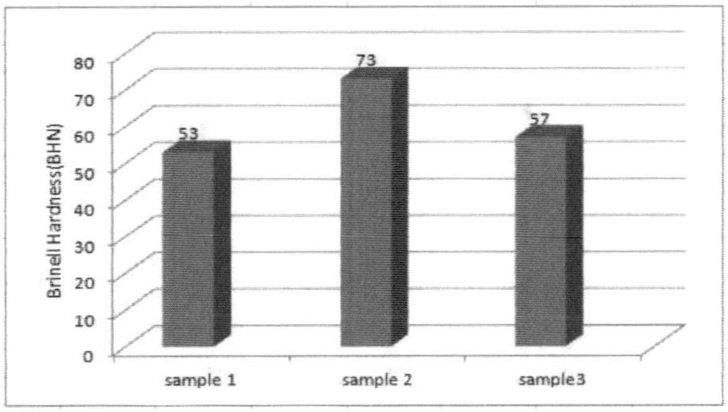

2(A) Result of Hardness Test- Graph

2(B) Result of Brinell Hardness Test- Specimen

5.3 Flexural Properties

Fig. 3 shows the result of Flexural Test, from which it is concluded that the sample which contains 1.5% of Silicon Carbide and 5% of copper is greater than the other samples. This is due to the presence of Silicon Carbide and Copper, which resist the bending property of a composite as compared to the other samples.

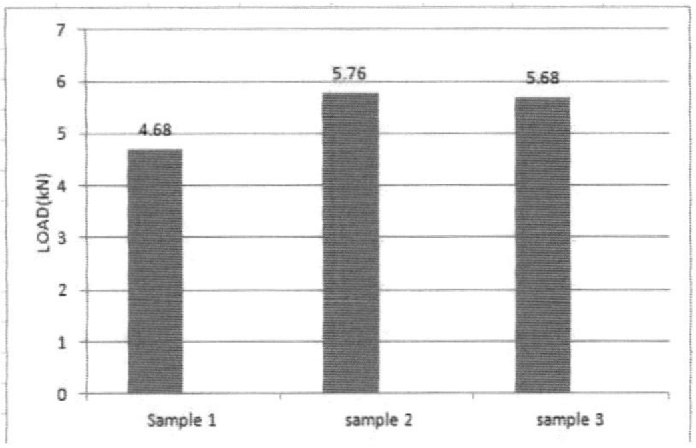

3(A) Result of Flexural Test-Graph

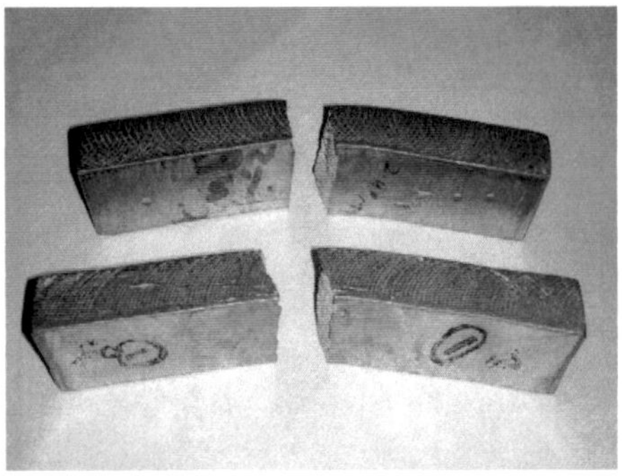

3(B) Result of Flexural Test- Specimen

6 Conclusion

In this work the tensile, flexural and hardness of Al metal matrix composites reinforced with Silicon Carbide and Copper were investigated.

It was found those samples which contain 1.5% of SiC and 5% Copper has more tensile strength and hardness than the rest of the samples. This may be attributed to the presence of less percentage of Silicon carbide which is balanced by the presence of Copper.

In Flexural (bending) test the sample (Sample 2) which contains 1.5% Sic and 5% Cu has more flexural ability compared to other samples. This may be due to the presence of copper which resists bending and balances the brittleness invoked by the presence of Silicon Carbide.

7 References

1. T.S Mohan kumar, S. Srinivas, M Ramachandra, K.V Mahendra and Madeva Nagaral. Effect of Fly ash and sic Particulates Addition on Mechanical Properties of Al -5wt.% Cu Alloy Composites. *IOSR-JMCE*. **Vol12, issue4, PP 01-05,** (2015)
2. H.C. Anilkumar, H.S. Hebbar and K.S. Ravishankar Mechanical Properties Of Fly Ash Reinforced Aluminium Alloy (Al6061) Composites. *IJMME*. **Vol.6, No.1, PP 41-45.** (2011)
3. Itskos G, Moutatsou A, Rohatgi P, Koukouzas N, Vasilatos C, Katsika E. Compaction of High-Ca Fly Ash-Al- and Al-Alloy-Composites: Evaluation of their Microstructure and Tribological Performance. *Coal Combustion and Gasification Products*. **Vol.3, PP 75-82** (2011)
4. Shivaprakash.Y.M, Yadavalli Basavaraj, K.V.Sreenivasa Prasad Comparative study of tribological characteristics of Al2024+10% fly ash composite in non-heat treated and Heat treated conditions. *International Journal of Research in Engineering and* Technology. , **Vol2, Issue11**. (2013)
5. Arun Kumar M. B. and R. P. Swamy Evaluation of Mechanical Properties of Al6061, Flyash and E-Glass Fiber Reinforced Hybrid Metal Matrix Composites. *ARPN Journal of Engineering and Applied Sciences*. , **VOL6, NO.5.** (2011)
6. Vivekananthan M and Senthamarai K Experimental Evaluation of Aluminium-Fly Ash Composite Material to Increase the Mechanical &Wear Behaviour by Stir Casting Method. *CARE Journal of Applied Research*. **(ISSN 2321-4090)** (2013)
7. K.V. Mahendra and K. Radhakrishna Fabrication of Al–4.5% Cu alloy with fly ash metal matrix composites and its characterization. *Materials Science-Poland*. , **vol. 25, pp. 57-68.**
8. Surendra Kumar Patel, Raman Nateriya, G. Dixit Microstructural Analysis, Microhardness and Compressive Behaviour of Dual Reinforced Particles Adc-12 Alloy Composite. *International Journal of Engineering Technologies and Management Research*. , **Vol.1, Issue1, PP 23-34** (2015)
9. Ramanpreet singh and Rahul Singla Tribological Characterization of Aluminium-Silicon Carbide Composite Prepared by Mechanical Alloying. *International Journal of Applied Engineering Research*. **Vol.7, No.11**. (2012)
10. Ashok Atulker, Kartikeya Tripathi and Vinod Pare Wear analysis of Aluminium Alloy for automobile piston against cast iron for cylinder liner. *12th National Conference on Machines and Mechanisms* (NaCoMM**-2005) PP 281-283**
11. Vinod Kumar, Phool kumar and Surender Sharma Wear Characterization of Aluminium/Sic/Al2O3 Hybrid Metal Matrix Composite Using Taguchi Technique. *International Journal of Innovative Research & Development* **Vol. 3 Issue 1,PP 34-40** (2014)
12. Gaurav Mahajan, Nikhil Karve, Uday Patil, P. Kuppan and K. Venkatesan Analysis of Microstructure, Hardness and Wear of Al-SiC-TiB_2 Hybrid Metal Matrix Composite. *Indian Journal of Science and Technology*. **Vol 8(S2), PP 101-105.** (2015)
13. L.Francis Xavier and Paramasivam Suresh Wear Behavior of Aluminium Metal Matrix Composite Prepared from Industrial Waste. *The Scientific World Journal*, **Vol. 2016, Article ID 6538345, 8 pages**, (2016)
14. Siddique Ahmed Ghias B.Vijaya Ramnath, Investigation of Tensile Property of Aluminium SiC Metal Matrix, *International Journal of Applied Mechanics and Materials*. **Volume 766-767, PP 252-256.** (2015)
15. P.Subramaniya Reddy, R.Kesavan, and B.Vijaya Ramnath, "Investigation of Mechanical Properties of Aluminium6061-Silicon Carbide, Boron Carbide Metal Matrix Composite ", Silicon, **Volume 1**, **PP 1-8**, (2016)
16. K. L. Meena, Dr. A. Manna, Dr. S. S. Banwait, Dr. Jaswanti an Analysis of Mechanical Properties of the Developed Al/SiC-MMC's. *American Journal of Mechanical Engineering*, **Vol. 1**, **No. 1**, **PP 14-19.** (2013)
17. Omkar Pansare, S.G.Kulkarni, S.S.Kulkarni. Statistical analysis by Taguchi method for buckling of aluminium matrix composite plate manufactured using Stir Casting, *International Journal of Science, Engineering and Technology Research*, **Vol. 4, Issue.7.** (2015)
18. B. Vijaya Ramnath, C.Parswajinan,C. Elanchezhian, S.V. Pragadeesh,C.Kavin,P.,R. Ramkishore, V. Sabarish, Experimental Investigation of Compression and Chemical Properties of

Aluminium Nano Composite, *International Journal Applied Mechanics and Materials*, **Volume 680, PP 7-10**. (2014)

19. B.Vijaya Ramnath, C. Elanchezhian, RM. Annamalai, S.Aravind, T. Sri Ananda Atreya, V. Vignesh, C.Subramanian, Aluminium metal matrix composites – a review, *Reviews on advanced material science*, **Volume 38**, **PP 55-60.** (2014)
20. B.Vijaya Ramnath, C. Elanchezhian, M.Jaivignesh, S. Rajesh, C. Parswajinan, A. Siddique Ahmed Ghias, Evaluation of Mechanical Properties of Aluminium Alloy-Alumina-Boron Carbide Metal Matrix Composites. *Materials and design*, **Volume 58, PP 332-338**. (2014)
21. N.Radhika, R. Subramanian, S. Venkat Prasat Tribological Behaviour of Aluminium/Alumina/Graphite Hybrid Metal Matrix Composite Using Taguchi's Techniques. *Journal of Minerals & Materials Characterization & Engineering*, **Vol. 10, No.5**, **PP.427**-433, (2011)
22. Manoj Singla, D. Deepak Dwivedi, Lakhvir Singh and Vikas Chawla Development of Aluminium Based Silicon Carbide Particulate Metal Matrix Composite. *Journal of Minerals & Materials Characterization & Engineering,* **Vol. 8, No.6**, **PP 455-467**, (2009).

Effect of Cryogenic Treatment on Micro-structure and Properties of Different Polymer Materials

Hongjun Ni[1,2], Lu Zhuo[1,2], Hai Gu[2], Shuaishuai Lv[1,2], Yu Zhu[1,2], Chongshi Shi[1,2], Xingxing Wang[1,2]

[1]*School of Mechanical Engineering, Nantong University, Nantong Jiangsu 226019,China*
[2]*Jiangsu Key Laboratory of 3D Printing Equipment and Application Technology,Nantong Institute of Technology, Nantong, Jiangsu 226002, China*

Abstract. In order to improve the mechanical properties of the polymer materials, three kinds of samples with the same size were manufactured by the 3D printing equipment. And the materials were polycarbonate, ABS and polypropylene, respectively. After -190°C+8h cryogenic treatment, the mechanical properties such as wear resistance and hardness were tested and the functional groups and internal molecule chain changes were observed by Fourier transform infrared spectrometer. The results showed that the mechanical properties of polymer materials can be improved effectively after the cryogenic treatment. Compared with the untreated samples, the hardness of the PC, ABS and PP samples increased by 27.6%, 10.8% and 20.3% respectively. Among the three materials, the wear resistance of ABS material was improved obviously.

1 Introduction

3D printing technology, known as additive manufacturing technology, is a manufacturing method based on three-dimensional models, making the adhesive materials stack layer by layer in the principle of discretion and accumulation. As an advanced manufacturing technology, 3D printing provides a new method of manufacturing materials and structures. And it is an important complement to the traditional manufacturing technology system[1].

At present Fused Deposition Modeling (FDM) has been widely applied in the field of science, education and family[2]. The thermoplastic polymer materials such as Polycarbonate (PC), Acrylonitrile - Butadiene - Styrene (ABS) and Polypropylene (PP) are used as the raw materials in this kind of technology. By the means of hot extrusion, the material is ordered stacked under the control of computer and finally the three-dimensional parts are obtained[3]. The 3D printed parts must have adequate mechanical strength, stiffness and wear resistance to ensure the durability of the final product[4]. Therefore, improving the performance of printed materials is the prerequisite for the universal application of 3D printed parts.

Cryogenic treatment is a kind of special heat treatment method. It usually refers to the cold treatment of raw materials in the environment of -190℃ to -230℃, so as to improve the microstructure and properties of the materials[5]. Related studies showed that cryogenic treatment can improve the mechanical properties such as hardness and wear resistance of metal materials[6], mainly due to the elimination of residual austenite and the initiation of nucleation sites for the subsequent precipitation of large number of very fine carbide particles[7]. However, the effect of cryogenic treatment on the properties of polymer materials need to be further explored[8].

In this paper, the three samples of different materials were produced by the technology of FDM. The samples were placed in -190℃ environment for 8h and then cooled to room temperature.The cryogenic treatment and untreated specimens were tested for wear resistance, hardness and Fourier transform infrared spectroscopy (FT-IR) to investigate the effects of cryogenic treatment on the properties of different polymer materials.

2 Experimental

The samples were tested by standards (GB/T 228.1-2010) of polymer materials mechanical performance, which were printed by rapid prototyping machines. The material of the the sample was PC, ABS and PP.

The samples of 150mm×4mm×20mm were drawn by 3D software[9]. Each of the three materials was divided into four groups and placed in SLX-80 cryogenic treatment box at -190℃ for 8h, respectively.

The wear resistance of the samples were tested by the MP-2B grinding and polishing machine (Laizhou Experimental Instrument Manufacturing Co., Ltd.) and electronic scales. The samples were exerted pressure of 1N and rotating at the speed of 500r/min with a 200-mesh electrostatic sand-grained alumina sand paper for 3min. The hardness of samples were tested by the HR-150A

Rockwell hardness taster. The German company Brook Vertex 70 Fourier infrared spectrometer scans range from 400cm-1 to 4000cm-1 and the resolution of is 1 cm-1, which was used to detect the change of the functional group and molecular chains.

3 Results and discussion

3.1 Mechanical properties

3.1.1 Wear resistance

The changes of the qualities of the samples were tested by the wear experiments for 3 min and listed in Table 1. It can be seen from the table that the wear resistance of the three materials is improved after the cryogenic treatment, especially the material of ABS. The cryogenic treatment can effectively improve the wear resistance of polymer materials with the poor properties. It is because the increased crystallinity can promote the interaction between molecules and the arrangement of material molecular chain tighter and more sequential as a whole.

Table 1. Test of wear resistance of different materials.

Experimental group	Weight before experiment (g)	Weight after experiment (g)	Loss of weight (g)	Proportion (%)
Untreated PC	4.1789	4.1690	0.0099	0.24
-190℃ for 8h PC	3.3038	3.3017	0.0021	0.06
Untreated ABS	2.4574	2.4257	0.0317	1.29
-190℃ for 8h ABS	2.7795	2.7631	0.0164	0.59
Untreated PP	2.4586	2.4479	0.0107	0.43
-190℃ for 8h PP	2.7194	2.7181	0.0013	0.05

3.1.2 Hardness

The hardness changes of the materials after cryogenic treatment were tested by Rockwell hardness tester. The surface of the samples were dotted repeatedly to eliminate the influence of extreme value. The results are shown in Table 2.

After cryogenic treatment, the hardness of the three kinds of materials increased by 27.6%, 10.8% and 20.3%, respectively.It can be seen that after the cryogenic treatment, the hardness of the three materials has been improved a lot. To some extent, this is because the development of the cryogenic treatment eliminated part of the residual stress.

Table 2. Test of hardness of different materials.

Experimental group	PC	ABS	PP
Untreated sample dot1	77	76.2	79
Untreated sample dot2	76	76.3	80
Untreated sample dot3	76	77	79
-190℃ for 8h sample dot1	79.5	77	81.2
-190℃ for 8h sample dot2	78	77.5	80.5
-190℃ for 8h sample dot3	78	77.5	81.2

3.2 Fourier transform infrared spectroscopy

The samples were characterized by Fourier transform infrared spectroscopy (Vertex 70) to observe the change of their functional groups [10]. The infrared spectra of the three kinds of materials are shown in Fig. 1.

There are three peaks near 1600, 1500 and 1450cm^{-1}, especially the first two zone is an important feature of the aromatic ring belt. As is shown in Fig.1, the peaks were produced by the stretch vibration of the aromatic ring skeleton C=C. The area of 1900-2500cm^{-1} is the cumulative double bond stretching vibration region, while the carboxyl may be at 3600cm^{-1}. After cryogenic treatment, the infrared transmittance spectrum of the three materials showed an increase in the peak transmissivity, indicating a decrease in absorption, which caused the properties of the functional groups weakened, the molecular chain length increased and the crystallinity increased [11-13]. And the arrangement of material molecular chain became more ordered, the stability got enhanced and the relevant mechanical properties were improved.

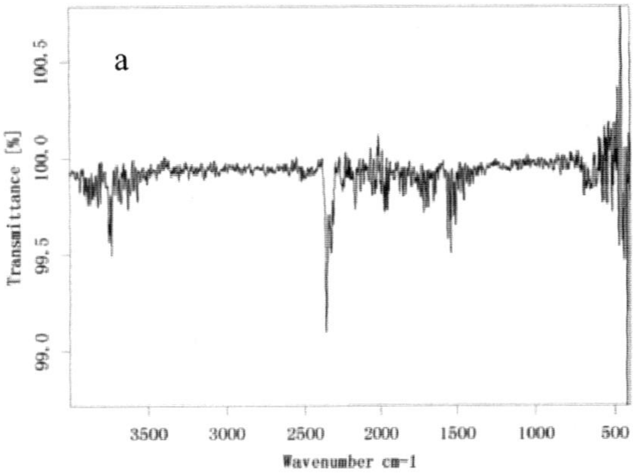

https://doi.org/10.1051/matecconf/201816601002

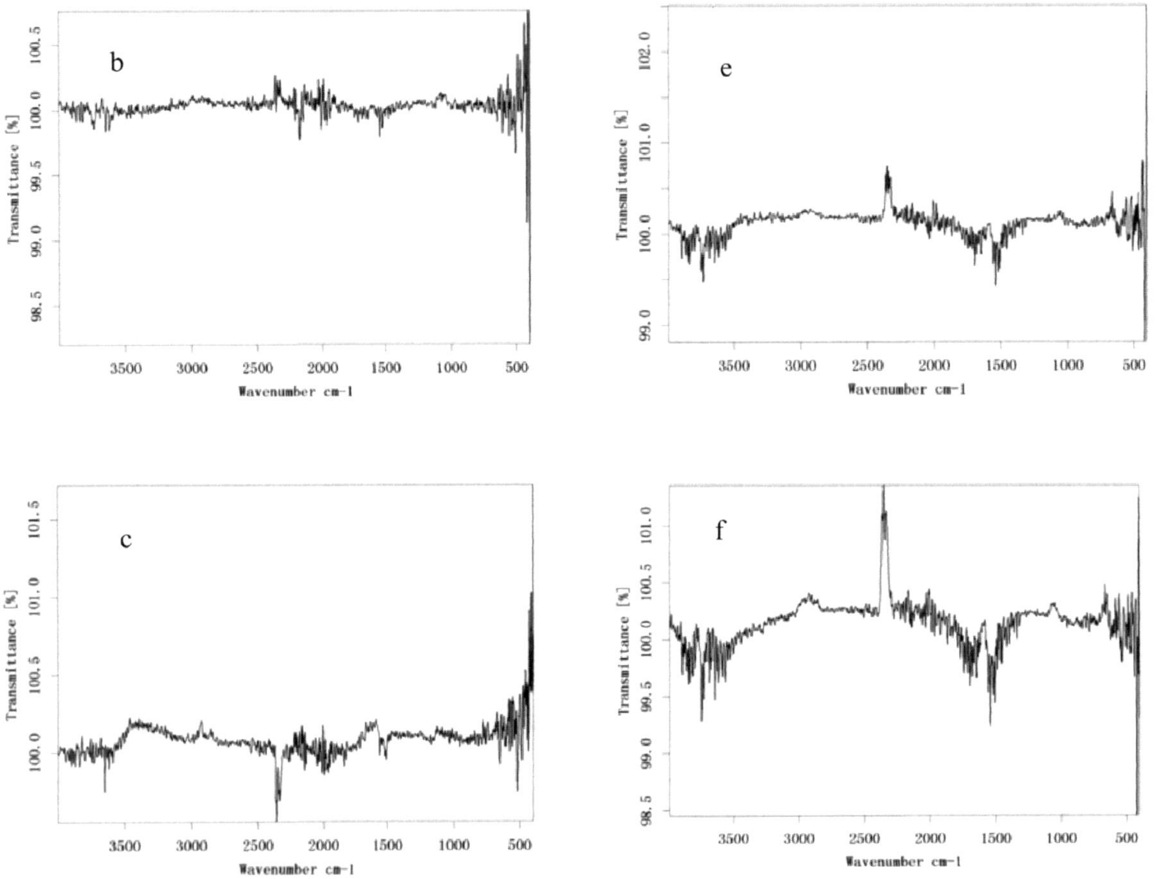

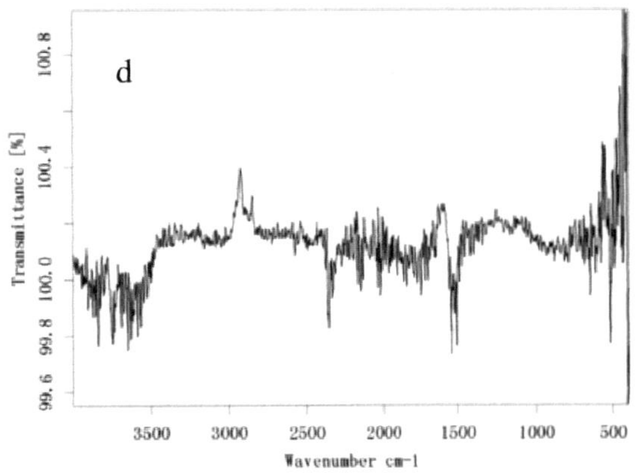

Fig. 1 Infrared Spectra of different samples before and after cryogenic treatment. (a) untreated PC (b) cryogenic treatment PC (c) untreated ABS (d) after cryogenic treatment ABS (e) untreated PP (f) after cryogenic treatment PP

4 Conclusion

(1) Cryogenic treatment can significantly improve the hardness and wear resistance of the material and the performance of materials with poor wear resistance can be greatly improved.

(2) Cryogenic treatment can improve the hardness of the material, and the hardness of the PC, ABS and PP samples increase by 27.6%, 10.8% and 20.3% respectively.

(3) After cryogenic treatment, the properties of functional groups are weakened, the length of molecular chain is increased, the crystallinity is increased, the arrangement is more orderly, and the stability of the material itself is enhanced, which can contribute to the promotion of its application

Acknowledgments

This work was financially supported by a project funded by the Priority Academic Program Development of Jiangsu Higher Education Institutions (PAPD), the Nantong City Applied Research Projects (GY12016054, GY12016056), the Natural Science Foundation of Jiangsu Province (No. BK20161289), Open Fund Project of Jiangsu Key Laboratory of 3D Printing Equipment and Application Technology (Nantong Institute of Technology)(2017KFKT02 、 2017KFKT09 、 2017KF KT10), and Nantong Key Laboratory of 3D printing technology and Application (CP12016002、).

References

1. Zhang X J, Tang S Y, Zhao H Y, Guo S Q, Li N, Sun B B, Chen B Q, J MATER ENG PERFORM, **44 (2)**, 122 (2016).
2. Xing H Y, Zou B, Li S S, Fu X S, CERAM INT, **43 (18)**, 16340 (2017).
3. Yu D M, Fang A, Zhang J B, Metal World, **6**, 6 (2013).
4. Gnanasekaran K, Heijmans T, Bennekom S V, Woldhuis H, Wijnia S, With G, Friedrich H, Applied Materialstoday, **9**, 21 (2017).
5. Li J, Zhou J Z, Xu S, Sheng J, Huang S, Sun Y H, Sun Q, Boateng E A, Materials Science and Engineering: A, **707**, 612 (2017)
6. Jagtar S, Pal S L, Ankur K, Friction & Wear Research, **1(2)**, 22 (2013).
7. Firouzdor V, Nejati E, Khomamizadeh F, J MATER PROCESS TECH, **206 (1 − 3)**, 467 (2008).
8. Liu H M, Wang J J, Li H H, Yao Y, China Plastics Industry, **45 (6)**, 1 (2017).
9. Wang X X, Lv Y, Lv S S, Liu H M, Ni H J,. EMERG MATER RES. (to be published)
10. Chen D, Liu F, Teng J, HEAT TREAT MET-UK, **33 (9)**, 66 (2010).
11. Rosenzweig D H, Carelli E, Steffen T, Jarzem P, Haglund L, INT J MOL SCI, **16 (7)**, 15118 (2015).
12. Sun J L, Trimby P W, Yan F K, Liao X Z, Tao N R, Wang J T, SCRIPTA MATER, **69 (5)**, 428 (2013).
13. Feng Q P, Deng Y H, Xiao H M, Liu Y, Qu C B, Zhao Y, Fu S Y, COMPOS SCI TECHNOL, **104**, 59 (2014).

Structure Optimization of Microbial Fuel Cell Reactor Based on Reverse Engineering and Rapid Prototyping

Shuaishuai Lv[1,2], Yangyang Zhu[1,2], Hai Gu[2], Hongjun Ni[1,2] ,Yue Meng[1,2] and Xingxing Wang[1,2]

[1]*School of Mechanical Engineering, Nantong University, Nantong Jiangsu 226019,China*
[2]*Jiangsu Key Laboratory of 3D Printing Equipment and Application Technology, Nantong Institute of Technology, Nantong, Jiangsu 226002, PR China,*

Abstract. The structure of Microbial Fuel Cell (MFC) reactor was analyzed and improved by Reverse Engineering (RE) and Rapid Prototyping (RP) technology. Points cloud data of MFC reactor was accessed by hand-held laser scanner. The space surface and entity model were reconstructed accurately, and the structure of the reactor was optimized based on Imageware and Solidworks software. The reactor model was manufactured by RP machine. The optimization efficiency of MFC reactor was improved based on the combination of RE and RP, which has a good reference value for the development of MFC technology and products.

1 Introduction

Microbial fuel cells (MFCs) have emerged as a sustainable technology for wastewater treatment that has potential in recycle bioelectricity from livestock wastewater, which use microbes as catalysts instead of expensive chemical catalysts compared to conventional fuel cells (FC) [1]. However, the current power generation of MFC has not reached the actual application level, and most of the researchs on MFC are also carried out in the laboratory. The performance of MFC is affected by the structure of the reactor to a great extent and optimizing the reactor structure is one of the most important ways to improve the performance of MFC [2]. With the rapid development of new energy vehicles, update cycle of FCs is shortened, and the complicated space structure of MFC still needs to be perfected at present, of which the production engineering can not meet the requirements of its development.

Reverse engineering (RE) is a redesign process to parameterize the known objects and reconstruct the model accurately [3,4]. RE pays more attention to the optimization redesign process than the imitation engineering. Rapid prototyping (RP) technology is the rapid model design and manufacturing technology based on digital technology. The combination of RE and RP can greatly improve the efficiency of product optimization [5].

In this study, RE and RP were utilized to optimize the MFC reactor, of which points cloud data was gained by hand-held laser scanner and processed by Imageware software. 3D model was created and optimized by Solidworks and printed by RP machine.

2 Parameterized measurement

Parametric measurement is to collect the coordinate points cloud data of the object space structure by means of professional measuring equipment to parameterize the object surface. Parametric measurement includes contact measurement and non-contact measurement, in which contact measurement is of high accuracy and adaptability but time consuming and easy to destroy the model, while non-contact measurement is not accurate but efficient. According to the space curved structure characteristics of the double-chambered MFC reactor (Figure 1), the EXASCAN™ hand-held laser scanner of Creaform company was selected [6]. Before the scanning, the imaging agent was sprayed on the surface of the bottle body and the fluorescent mark points were added according to the reactor structure (as shown in Figure 2). The destination fileof scanning was saved as .stl format after initial treatment.

Figure 1. MFC reactor

Figure 2. The scanning of MFC reactor

3 Parametric design

Parametric design is to reprocess the parameterized measurement data based on the general idea of point-curve-surface-entity, including pre-processing of points cloud, reconstruction of surface model and 3D solid modeling. Firstly, the redundant points generated in the scanning process were removed, and the number of the original points cloud was reduced from 16202 to 6237 using range sampling method. Based on the step by step localization method, the points cloud data of the bottle body was modified in sequence according to the characteristic relation, as shown in Figure 3.

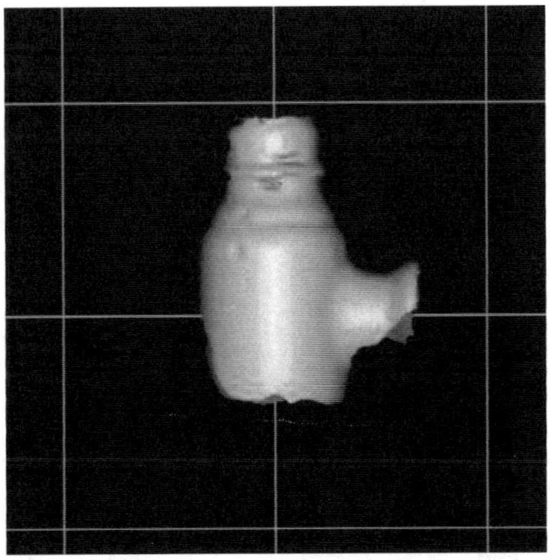

Figure 3. Position correction of points cloud data

Surface reconstruction is the process of fitting the surface based on the regularities of points cloud distribution according to the specific algorithm, in which the characteristic curve is the basis of regular surface modeling, and the characteristic curve can be used to make up the defects of the solid parts according to the design requirements to optimize product. According to the characteristic points cloud distribution fitting Non-Uniform Rational B-Splines (NURBS) surface, which is the most widely used technology of complex surface modeling, and improving the quality of NURBS surface effectively to meet the design requirements withthe modification of the control vertex. The points cloud data of the reactor bottle body was divided into four parts: the bottle body, the bottle interface, the bottleneck and the transition surface. Firstly, the points cloud of bottle body was selected and the characteristic points cloud was intercepted (Figure 4a), then the tolerance curve was constructed by points cloud and swept into the surface, and trimmed according to the actual model, as shown in Figure 4b. The processing method of the bottle interface and the bottleneck surface was the same as the bottle body's. The deviation in fitting surface was accurately adjusted with the aid of NURBS surface because the structure data of the transition surface of three rotating surfaces was complex, and the surfaces were seamlessly connected (Figure 5). Error distribution between final reconstructed surface and points cloud was analyzed as shown in Table 1., and the maximum surface deviation was 0.2734mm.

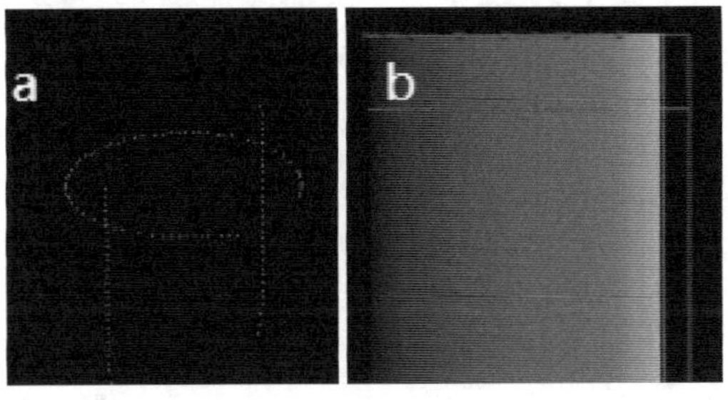

Figure 4. Surface reconstruction of bottle body

a. Interception of characteristic points cloud; **b.** Trimming of swept surface

Figure 5. Reactor surface

Table 1. Error distribution between reconstructed surface and points cloud

Deviation	Maximal Value /mm	Average Value /mm	Standard Value /mm
Lateral deviation	0.0000	0.0000	0.0000
Negative normal deviation	-0.2734	-0.0564	0.0475
Geometric deviation	0.2734	0.0531	0.0460
Normal deviation	0.1276	0.0225	0.0384

The processed surface data was imported into Solidworks for 3D solid modeling, and the model was substantiated and extracted (Figure 6). The points cloud data of interface groove structure and bottleneck screw structure was not a precise measurement because of their small complex structure, which needs the utilization of Solidworks to further improve the model according to the actual measurement or design requirements (Figure 7). Because the bottle cap was complex and not of imposing stature, the 3D model was constructed directly with the aid of Soldiworks according to the design requirements (Figure 8). The final assembly diagram of the reactor was shown as Figure 9.

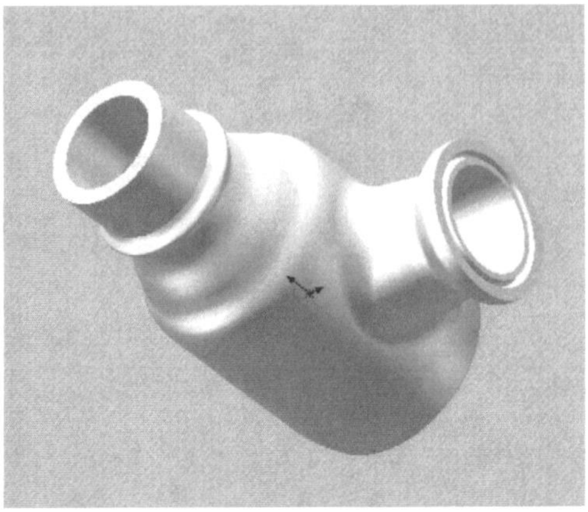

Figure 6. Reactor shell extraction

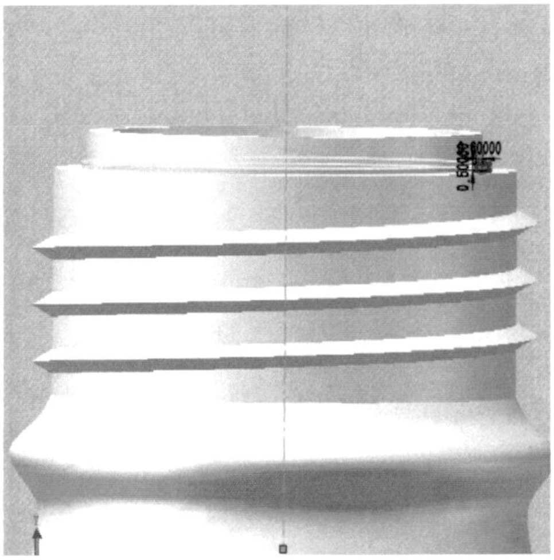

Figure 7. Sketch diagram of thread machining

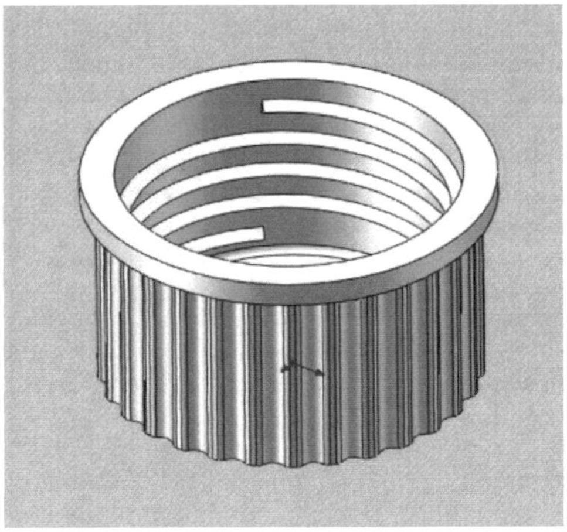

Figure 8. Bottle cap of reactor

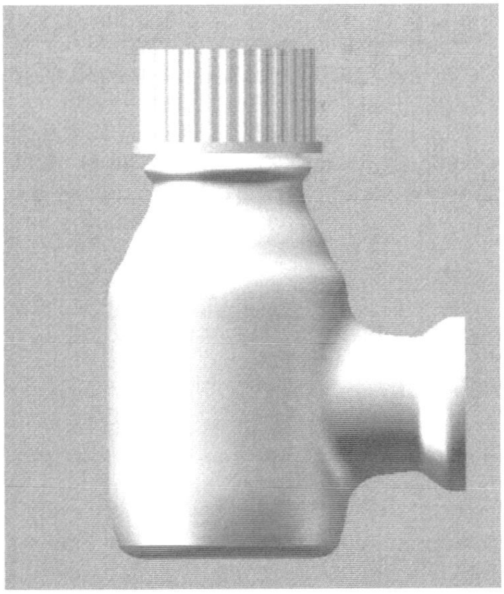

Figure 9. Reactor assembly drawing

According to the experimental operation, the phenomenon of water leakage was easy to occur around the bottle interface part of the original reactor when connected. In order to eliminate the influence of the experimental operation error on the experimental results, the groove structure of the original interface was changed into the flange shape and fixed with the threaded connection to optimize the structure of the reactor [7].

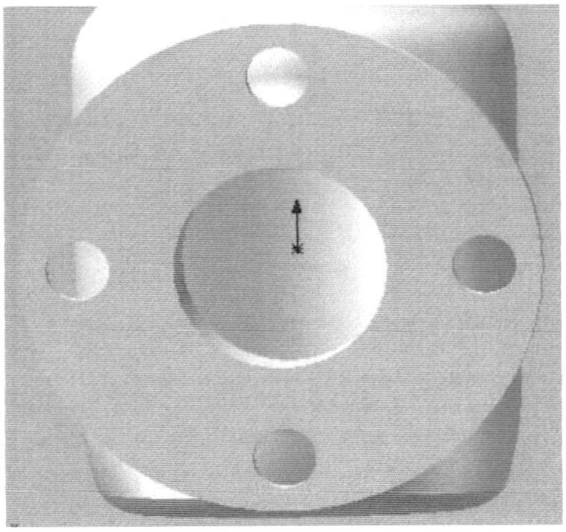

Figure 10. Flange plate of reactor

4 Rapid Prototyping

At present, the RP technology is to layer 3D solid model using 2D slicing method, by which obtained the section contour of each layer, in order to achieve the increase in material plane along the direction of material increase manufacturing technology [8-10]. In this study, the 3D modeling data of model was imported into the MEM320A melt extrusion rapid prototype, and the solid manufacturing of the MFC reactor was carried out based on ABS S301 and ABS B601 materials.

Entity processes for MFC reactor are as follows:

(1) Preheating. Before the RP manufacturing, preheating treatment was needed to ensure that the temperature of the molding chamber reached 60℃, the molding material reached 260℃ and the supporting material reached 240℃.

(2) System initialization. Initializing the system and ensuring the safety of equipment connection.

(3) Model loading. Converting 3D model data into .stl format and importing into software, then coordinating appropriate coordinate system.

(4) Debugging. Detection and correction of model errors.

(5) Table adjustment: Setting the height of the table and the distance between the nozzle and the worktable to improve the printing efficiency.

(6) Parameter setting: Adjusting printing parameters, such as precision, support material, etc.

(7) Start printing: The machine starts automatically and stopping after the model print was finished.

Figure 11. RP entity of MFC reactor

5 Conclusions

Reverse engineering software (Imageware) and forward engineering software (Solidworks) were combined to reverse reconstruct the 3D model of the MFC reactor with complex spatial structure, which greatly shorten the production cycle without the traditional engineering drawing.

The effective combination of RE and RP was in line with the requirements of MFC development, and provided reference for optimizing the MFC reactor.

Acknowledgments

This work was financially supported by a project funded by the Priority Academic Program Development of Jiangsu Higher Education Institutions (PAPD), Natural Science Foundation of Jiangsu Province (No. BK20161289), Applied Research Projects of Nantong City (GY12016054, GY12016056), Open Fund Project of Jiangsu Key Laboratory of 3D Printing Equipment and Application Technology (Nantong Institute of Technology)

(2017KFKT02 、 2017KFKT09 、 2017KFKT10), and Nantong Key Laboratory of 3D printing technology and Application (CP12016002).

References

1. L. Zhuo, X.X. Wang, S.S. Lv, M.Y. Huang, H.J. Ning, Mod Chem Ind **8**, 41 (2017)
2. J. Ma, D.Y. Su, H.K. Pu, X.X. Wang, Q.Q.Chen, Chinese Journal of Power Sources **39 (10)**, 2318 (2015)
3. Y. Zhu, X.W. Li, J.D. Wei, X.X. Wang, Y.Y. Zhu, H.J. Ni, Plastics Science and Technology **45 (4)**, 79 (2017)
4. X.X. Wang, J.D. Wei, Y. Pei, Y. Zhu, H.J. Ni, Appl. Mech. & Mater. **752-753**, 1301 (2015)
5. T. Rayna, L. Striukova. Technological Forecasting & Social Change **102**, 214 (2016)
6. D.T. Li, X.B. Feng, P. Liao, H.J. Ni, Y.D. Zhou, M.Y. Huang, Z.Y. Li, Y. Zhu, Lecture Notes in Electrical Engineering **269** , 1919 (2014)
7. C. J, C.S. Xiang, R.Z. Liu, Opto-Elec Eng **43 (7)**, 40 (2016)
8. V.N. Chougule,A.V. Mulay, B.B. Ahuja, Procedia Eng. **97**, 212 (2014)
9. K. Bian, K.E. Yinglin, Chin. J. Mech. Eng. **26**,197 (2013)
10. S.A. Khaled, J.C. Burley, M.R. Alexander, J. Yang, Int. J. Pharm **494 (2)**, 643 (2015)

Comparison of Ortho-planar Spring design optimization based on Linear Elastic and Hyper Elastic Materials

Ruetai Graipaspong[1] and Teeranoot Chanthasopeephan[1]

[1] *Department of Mechanical Engineering, King Mongkut's University of Technology Thonburi, Bangkok, 10140 Thailand*

Abstract. In this paper, compliant Ortho-planar spring was designed based on a three-dimensional topology optimization method. The computation was developed using MATLAB programming. The objective of this work was to apply dual method to design an Ortho-planar spring while the design should have minimum mass and at the same time satisfy a set of constrained displacement. Throughout this paper, we analyzed a method for designing an Ortho-planar spring using linear elastic material and hyperelastic material. The results showed that under small displacement conditions, the output displacement, maximum stress magnitude, and the maximum stress of linear elastic assumption and hyper-elastic material were relatively close to each other. However, the mass fraction and the layout as the result of the optimization process was different. As for larger displacement, the maximum stress of linear elastic material appeared 2.59 times higher than the maximum stress of the hyper-elastic material model. The topology optimization output based on linear material was invalid because the topology of the computed Ortho-planar spring was not appeared as a one-piece layout while the design based on nonlinear material looked promising.

1 Introduction

Ortho-planar spring in general is described as a mechanism that lifts the platform up from its plane. Compliant mechanism, on the other hand, is a mechanism moved by deflection of members. Instead of using joint to lift the platform, compliant Ortho-planar spring moves through the deflection of members [1, 2]. The avantages of compliant Ortho-planar springs are not only the ease of manufacturing process, but also that their compactness, resistance to wear, and light weight [2]. Multiple applications of Ortho-planar springs include valves, actuators, pneumatic controllers, etc. [3-5]. In each application, the layouts of the compliant Ortho-planar springs are designed differently.

Topology optimization is a mathematical approach that synthesizes the optimum layout of the structures within the given loads and boundary conditions. In the past, various calculation tools were used for topology optimization. Mathematica had been used to design Michell truss structures [6]. Aside from Mathematica, MATLAB was also used for calculation. Through MATLAB implementation, Sigmund [7] introduced 99-line code for 2-D topology optimization using optimum criteria method. Afterward, the 99-line code was shortened by Andreassen [8] until the code reached 88 lines. One advantage of decreasing lines of the codes was it took less time to synthesize the optimal layout. Although the 88-line MATLAB code successfully reduced computational time, the computation was limited to only two-dimensional design. As a result, Liu [9] had developed a three-dimensional topology optimization MATLAB code using optimum criteria method. The designs of Sigmund, Andreassen, and Liu were based on optimum criteria method, where the objective function was to minimize compliance while limited the volume fraction. The limitation of volume fraction was used as a condition while computing for the structure with the desired weight. Aside from using minimize compliance as the objective function, other objective functions could still be used. Sui [10] applied dual method for 2-dimensional topology optimization with the objective function of mass minimization instead of compliance minimization. Sui's constrained the displacement of the loaded structure, not the volume fraction, unlike the above mentioned works. By using different calculation method, different objective functions could be appropriately chosen for each case.

Although many topology optimization models worked well with linear finite element, it could not be used for large displacement analysis where nonlinearity plays an important role. Nonlinearity analysis usually suits with geometrically nonlinear, material nonlinear, or combined nonlinear problems. The analysis and design of nonlinear compliant mechanisms using hyperelastic materials was explored by Bruns [11]. Nonlinear hyperelastic material was required for large displacement compliant mechanisms. The formulated equations were solved by using Newton-Raphson method. Jung [12] discussed the differences between the optimal layouts and deformed shapes of linear design, materially nonlinear design, geometrically nonlinear design, and combined design.

Nonlinear material was not only modelled by using functions between stress and strain, but also from deriving constitutive equations in terms of second Piola Kirchoff stress and Green-Lagrange strain tensors. Topology optimization of compliant mechanisms based on nonlinear design using meshless models was done by Du [13]. Their method suited well for large deformation of a compliant mechanism design. Meshless models discretized the design domain by nodes, instead of elements. The proposed numerical optimization process could be used for finite element method as well.

In this paper, compliant Ortho-planar spring was designed based by developing a MATLAB code to compute for three-dimensional topology optimization. Dual method was applied and the objective function was to minimize mass. The constraint was a define set of displacement. The aim of this research was to analyze and compare the design of an Ortho-planar spring based on linear elastic material versus hyper-elastic material.

2 Topology optimization

The main steps of topology optimization are presented in the flowchart as shown in Figure 1. According to the flow chart, the steps toward topology optimization process in Figure 1 were given as follow. To design an Ortho-planar spring, the first step started with setting a specified set of length, width, and height of the prototype. Then, an objective function which subjected to its constraint was constructed. The initial conditions were set. To prevent checkerboard structure, mesh-independency filter was added to the process.

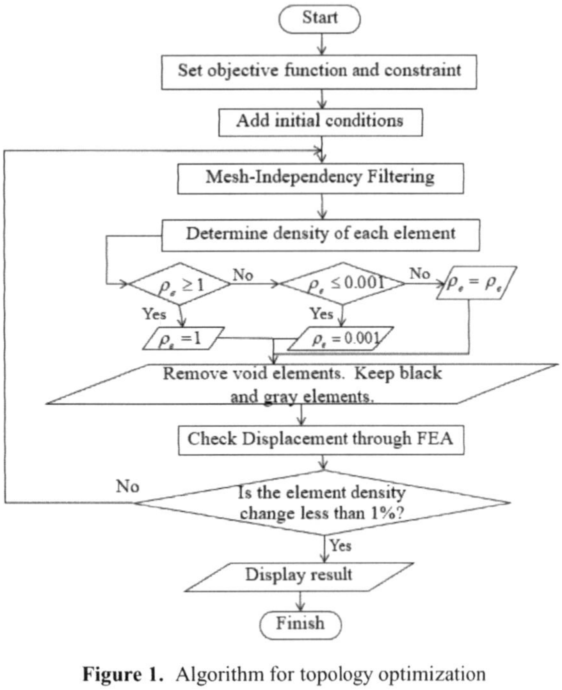

Figure 1. Algorithm for topology optimization

The following steps start with determine element density, remove elements that have element density of $\rho_{\min}$ (void elements), and keep the elements with element density in the range between $\rho_{\min}$ and 1 (gray and black elements). As the element density was calculated, the outcome of the displacement should be checked whether it has the same value as the constraint or not. Afterward, the element density of the current iteration is compared to the element density in previous iteration. If either changes in element density is higher than 1 percent or the output displacement contradicts the constraint, the process has to be repeated. Once the output displacement and change in element density satisfies the conditions, the calculation is finished.

In this paper, the objective function is to minimize mass, while constraining the displacement of Ortho-planar spring. The summation of element mass, m_e, lead to total mass, m, which depends on element density, ρ_e. The element density must be between minimum density and 1.

$$\begin{aligned} \min \quad & m(\rho) = \sum_{e=1}^{N} \rho_e (\rho v)_e = \sum_{e=1}^{N} \rho_e m_e \\ s.t. \quad & u(\rho) = \bar{u} \\ & 0 < \rho_{\min} \leq \rho_e \leq 1 \end{aligned} \tag{1}$$

According to the work of Sui [10], the structure displacement $u(\rho)$ is the sum of element displacement. From work-energy theorem and stiffness matrix displacement is obtained as

$$u = \sum_{e=1}^{N} \left(P_e^R\right)^T \left(K_e^0\right)^{-1} P_e^v \rho_e^{-p} \approx \sum_{e=1}^{N} \frac{D_e^0}{\rho_e^p} \tag{2}$$

where p and D_e^0 are the density penalty and displacement coefficient respectively. K_e^0 and P_e^v are the element stiffness matrix per unit element density and element nodal force associated to the virtual load respectively. The value of input penalty depends on the Poisson's ratio of the selected material. If the input penalty is not high enough, the solution may not reach its global optimum[14]. Therefore, the objective function is as following:

$$\begin{aligned} \min \quad & m(\rho) = \sum_{e=1}^{N} \rho_e (\rho v)_e = \sum_{e=1}^{N} \rho_e m_e \\ s.t. \quad & \sum_{e=1}^{N} \frac{D_e^0}{\rho_e^p} = \bar{u} \\ & 0 < \rho_{\min} \leq \rho_e \leq 1 \end{aligned} \tag{3}$$

Using a Lagrange multiplier to combine the equations and solving the Lagrangian equations, element density equation is then obtained.

$$\rho_e = \left(\frac{\sum_{k=1}^{N} \left(D_k^0\right)^{\frac{1}{p+1}} \left(m_k^0\right)^{\frac{1}{p+1}}}{\bar{u}} \right)^{\frac{1}{p}} \cdot \left(\frac{D_e^0}{m_e^0} \right)^{\frac{1}{p+1}} \quad (4)$$

The element density limit has been given between $\rho_{\min}$ and 1. Therefore, the element density is equal to 1, or element is solid, when the calculated density value exceeds 1. Moreover, it is given at the minimum value or as void, if the density from calculated from the equation is less than or equal to $\rho_{\min}$. The element is present but would have gray color as the calculated value is between $\rho_{\min}$ and 1.

3 Linear material and nonlinear hyperelastic material analysis

3.1. Linear Material Analysis

The difference between linear material and nonlinear material analysis is the stiffness matrix. With linear material assumption, the stiffness matrix was considered as a constant matrix. The element stiffness matrix was in the form of

$$Ke_{linear} = \int_V B^T C(E, \nu) B dV \quad (5)$$

where B and C were strain-displacement matrix and constitutive matrix respectively. In linear finite element analysis, the constitutive matrix was a function of Young's modulus (E) and Poisson's ratio (ν).

3.2 Nonlinear Hyperelastic Material Analysis

Equation(5) could not be used to determine the element stiffness matrix. In the case of nonlinear material with linear geometry assumption, the stiffness matrix depended on displacement. The tangent element stiffness matrix was described in the form of

$$Ke_{nonlinear} = \int_V B_m{}^T S B_m dV \quad (6)$$

where B_m was the strain-displacement matrix for nonlinear material. The S matrix was the second Piola Kirchoff stress matrix, in which the values depend on the displacement of the structure. Second Piola Kirchoff stress was calculated from the models that were used to fit the stress-strain curve. There were multiple models that could be used to fit the hyper-elastic material stress-strain curve, depending on each case. In this paper, Yeoh's model was applied.

$$W = c_1(I_1 - 3) + c_2(I_1 - 3)^2 + c_3(I_1 - 3)^3 \quad (7)$$

where W and I_1 are strain energy and first invariant respectively. Aside from that, c_1 , c_2 ,and c_3 were constants. From strain energy equation, second Piola Kirchoff stress was obtained by

$$S = c_1 I_{1,E} + 2c_2(I_1 - 3) I_{1,E} + 3c_3(I_1 - 3)^2 I_{1,E} \quad (8)$$

where $I_{1,E}$ was the derivative of first invariant.

As it was previously mentioned, the stiffness matrix of hyper-elastic materials depended on the displacement. Therefore, the finite element equation could not be solved directly. Various numerical methods have been mentioned to solve nonlinear finite element models. The most suitable method for this case was Newton-Raphson method [15, 16]. This strategy worked by using the tangent stiffness matrix " $Ke_{nonlinear}$ " and the residual force " R " to find the displacement of the iteration.

$$R(U_{x-1}) = [K_{nonlinear}(U_{x-1})](\mathrm{U}_{x-1}) - F \quad (9)$$

$$R(U_{x-1}) = [K_{nonlinear}(U_{x-1})](\Delta \mathrm{U}) \quad (10)$$

The displacement of the current iteration, x , was found to be $U_x = U_{x-1} + \Delta U$. Since residual force was the difference between internal and external force, the loop continued until the residual force reached zero.

4 Results and discussion

The topology design of Ortho-planar spring had been calculated as load "F" was applied in the center of a 50mm x 50mm x 1mm solid structure. Each element was a given size of 1mm x 1mm x 1mm. Thus, the number of element along the x, y, and z directions were 50, 1, and 50 respectively. Figure 2 showed that all four sides (blank surfaces) of the initial design were fixed, while the shaded surfaces represented free surfaces.

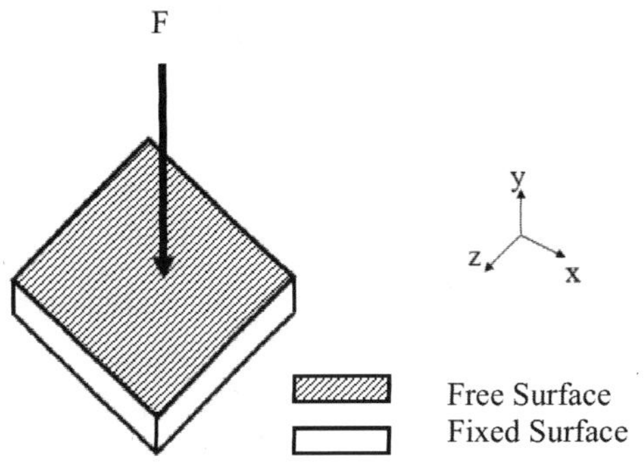

Figure 2. Initial condition of Ortho-planar spring

Table 1. Results from small displacement input

	Linear Elastic Material	**Hyper-Elastic Material**
Output Displacement (mm)	1.09	1.099
Maximum Stress (MPa)	9.210	11.115
Mass Fraction	0.3	0.25
Layout*		

*The circles represent points of maximum stress.

Table 2. Results from large displacement input

	Linear Elastic Material	**Hyper-Elastic Material**
Output Displacement (mm)	5.00	5.01
Maximum Stress (MPa)	28.835	11.126
Mass Fraction	0.1	0.07
Layout*		

*The circles represent points of maximum stress.

Table 1 and Table 2 presented the results from small displacement and large displacement of the linear and nonlinear materials. In Table 1, the displacement was limited to 1.1 mm. The maximum stress and the location of the maximum stress of linear elastic material and hyper-elastic material were close to each other. However, the outcome of mass fraction and layout were somewhat different.

In Table 2, the displacement was limited to 5 mm. Although the locations of maximum stress of both cases were close to each other, the maximum stress of linear elastic material was 2.59 times higher than the maximum stress as the result of the hyper-elastic material model. This was due to the behavior of the material; hyperelastic materials have different behavior as compared to the linear elastic material as the displacement increases. Aside from that, the layout of the linear elastic material did not come out as one piece layout. The fragmented layout was invalid. At large strain, it was recommended to use nonlinear material properties to design Ortho-planar spring.

5 Conclusion

The applications of Ortho-planar spring include valves, actuators, pneumatic controllers, etc. In this paper, an Ortho-planar spring was designed through a topology optimization. With the given boundary conditions, the comparison between the effects of linear material property and nonlinear material property on topology optimization had been analyzed.

First, the displacement constraint was set as 1.1 mm. The output displacement, maximum stress, and positions of maximum stress of linear elastic assumption and hyper-elastic material tend to be the same while the mass fraction and the layout were different. Then, the displacement constraint was set to 5 mm. The output displacement, mass fraction, and positions of maximum stress tend to be the same. With larger displacement, the maximum stress of linear elastic assumption was 2.59 times higher than the maximum stress of the hyper-elastic material model. This was due to the different between the behavior of materials. Hyperelastic materials have different behavior than linear elastic material as the displacement increases. Also, the layout based on the linear elastic material assumption did not come out as a one piece layout, which meant that the layout of linear material case was invalid. Therefore, it was recommended to use nonlinear material properties to design Ortho-planar spring as large deformation occurred.

Acknowledgement

The authors would like to express gratitude to Smart D-Vision Company Limited and the Research and Researchers for Industries (RRI) scholarship No. MSD56I0018 under Thailand Research Fund (TRF) for financial support. We also thank Mechanical Engineering department, King Mongkut's University of Technology Thonburi for providing the equipment for testing.

References

1. J. J. Parise, L. L. Howell and S. P. Magleby, Mechanism and Machine Theory **36** (11–12), 1281-1299 (2001).
2. L. L. Howell, *Compliant mechanisms*. (John Wiley & Sons, 2001).
3. B. D. Olivier Smal, Benoit Raucent, Michael De Volder, Jan Peirs, Dominick Reynaerts, Frederik Ceyssens, Johan Coosemans, Robert Puers, Journal of Micro-Nano Mechatronics **4** (3), 131-143 (2008).
4. O. Smal, B. Raucent, F. Ceyssens, R. Puers, M. Volder and D. Reynaerts, Micro-Assembly Technologies and Applications: IFIP TC5 WG5.5 Fourth International Precision Assembly Seminar (IPAS'2008) Chamonix, France February 10–13, 2008, 75-86 (2008).
5. B. R. a. H. J. Olivier Smal, International Journal for Simulation and Disciplinary Design Optimization **3** (2), 356 - 362 (2009).
6. T. Sokół, Structural and Multidisciplinary Optimization **43** (2), 181-190 (2011).
7. O. Sigmund, Struct. Multidiscip. Optim. **21** (2), 120-127 (2001).
8. E. Andreassen, A. Clausen, M. Schevenels, B. S. Lazarov and O. Sigmund, Structural and Multidisciplinary Optimization **43** (1), 1-16 (2011).
9. K. Liu and A. Tovar, Structural and Multidisciplinary Optimization **50** (6), 1175-1196 (2014).
10. G. Y. Yunkang Sui, in *10th World Congress on Structural and Multidisciplinary Optimization* (2013).
11. T. E. Bruns and D. A. Tortorelli, Computer Methods in Applied Mechanics and Engineering **190** (26), 3443-3459 (2001).
12. D. Jung and H. C. Gea, Finite Elements in Analysis and Design **40** (11), 1417-1427 (2004).
13. L. C. Yixian Du, International Journal of CAD/CAM **8** (1), 1-10 (2008).
14. M. P. B. O. Sigmund, *Topology optimization: theory, methods, and applications*. (Springer, 2003).
15. J. N. Reddy, *An Introduction to Nonlinear Finite Element Analysis*. (OUP Oxford, 2004).
16. N.-H. Kim, *Introduction to Nonlinear Finite Element Analysis*. (Springer US, 2015).

Improved Fractographic Strength Estimates Based on Surface Profilometry

Lingyue Ma[1] and Roberto Dugnani[1]

[1]*University of Michigan, Shanghai Jiao Tong University Joint Institute, UM-SJTU JI, 800 Dongchuan Road, Shanghai, 200240, China*

Abstract. Fractography is a valuable method that uses post-mortem topographical information to estimate the stress field near the fracture origin and help establish the root cause of failures. Typically, in glass and ceramics the mirror radius is one of the features sought for by fractographers since its length could be empirically related to the sample's strength. The mirror radius is usually subjectively estimated by fractographers though microscopy measurements. Nonetheless, variations in the estimates introduced by inconsistent viewing modes and the subjectivity of observers could lead to substantial errors even when standard protocols such as ASTM C1678 were followed. In this manuscript, a novel method combining a fracture mechanics model describing the mist formation in silicate glasses with profilometry data carried out by confocal laser scanning microscope is introduced. The new method was shown to be able to objectively establish the mirror-mist boundary. Furthermore, it was found that the proposed technique was repeatable within 2% regardless of the magnification or imaging mode used. Whereas the average strength estimated per ASTM C1678 by eight individual observers was influenced by both the magnification and the imaging mode used and displayed standard deviation of over 3%.

1 Introduction

Fractography is often used to estimate the fracture strength of brittle materials by analyzing revealing fracture patterns. In silicate glass and ceramics, these telltale surface marks include a relatively smooth region surrounding the fracture origin, named the "mirror region", a coarse, hazy "mist region" adjacent to mirror region, and a region with feather-like, non-coplanar segments called the "hackle region".

For over half a century, fractographers have been estimating the strength of the fractured samples by measuring the distance from the fracture origin to the boundary of the mirror/mist region. This characteristic length scale, known as the mirror radius or inner-mirror radius, could correlated to the strength of the material using Orr's equation [1]:

$$\sigma_f = \frac{A}{\sqrt{R}} + C. \qquad (1)$$

Where R is the mirror radius, A and C are linearly regressed constants, and σ_f is the fracture strength. The constant, A is often referred to as the mirror constant. Orr's original work did not include C, but its use has become standard especially for $C > 10\text{MPa}$, as recommended by both NIST and ASTM standards [2, 3].

The process of measuring mirror radius conventionally is subjective and hence it naturally leads to estimating error. In order to alleviate this problem both ASTM standard [2] and the NIST guide [3] introduced standard procedures for mirror radius measurements. Nonetheless, even when carefully following the directives outlined in the standard, the measuring process inevitably leads to variations due to the subjective nature of the process. Additional variations in the estimations are caused by factors such as the type of viewing method, the illumination used, and the magnification used [4].

Arguably, the major source of errors when investigating fractures is introduced by the lack of a clear definition of what constitutes the mirror/mist boundary [5]. Normally observers optically discern the mist as the relative rough surface where – based on their personal perception – the visible light begins to be scattered by the increasingly rougher surface. This suggests that the threshold of the roughness should be a fraction of the range given by the wavelength of visible light (i.e., 0.39-0.80 μm) [4], yet no clear demarcation actually exists and no clear arbitrary roughness threshold has been set by neither NIST nor ASTM standards. Consequently, it is expected that the mirror radius be affected by viewing methods and in general it was found that researchers using non-optical tools suggested lower roughness values to describe the mirror-mist boundary [6-9].

In this manuscript, a novel method is described to obtain accurate and repeatable estimations of the mirror radius. The method combines knowledge of the theoretical behaviour describing the mist formation in silicate glasses based on fracture mechanics principles with profilometry data obtained by confocal laser scanning microscope (CLSM). The novelty of the method consists in combining advanced visualization technique with the most current understanding of the brittle fracture behaviour.

2 Methods

In order to consistently establish the length of the mirror radius, a method combining the theoretical description of the mirror-mist boundary locus with profilometry data of the fracture surface was used. In this section, the numerical algorithm used to manipulate the fracture surface profilometry data was described in detail.

The profilometry scan of the fracture surfaces were carried out by CLSM (Zeiss LSM 700/Olympus PMG 3) with the operating parameters summarized in table 1. The methods to determine the two extremities defining the mirror radius, i.e. the origin, and the mirror-mist boundary were described separately.

Table 1. Operating parameters for CLSM.

Laser wavelength	Laser Intensity	Pinhole	Channel Gain	Digital offset
405nm	1.2	1AU[1]	~350	1.0

2.1 Samples preparation

Two sets of glass plates were tested by four-points bending tests (4PTB) according to ASTM standard C158-02 [10]. The first set included alumino-silicate glass of 0.7mm thick, and the second set included soda-lime glass of 2mm thick. The tests were conducted on a MTS testing system (model 45) with a 50kN load cell (resolution of 0.01N), outer/inner span of 200/100mm and rollers' diameter of 10mm. The fractured surfaces were first scanned by CLSM and then imaged by conventional optical microscopy both in bright-field and differential interference contrast (DIC).

2.2 Origin determination

According to ASTM standard [2], the mirror radius on fractures with origins away from the edges could be obtained as half the length between the two mirror-mist boundaries at the stress-free surface. Nonetheless, for samples with corner crack origins or near-the-edge origins, the length of the mirror radius is not as straightforward to obtain as it requires knowledge of the location of the fracture origin.

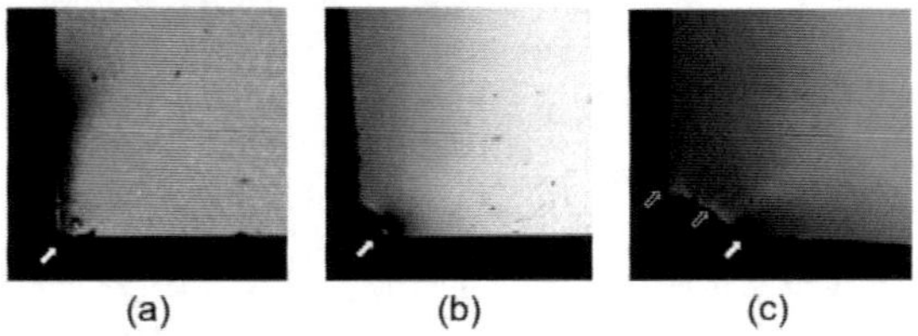

Figure 1. Origins under CLSM from different glass samples fractured by 4PTB.

[1] 1.01 Airy Units = 13.0μm section.

Fig. 1 shows various corner fracture origins from samples fractured by 4PTB with white arrows pointing at the origin. When the fracture origins were not very obvious when inspected by bright field, different illumination method such as DIC, or non-optical methods such as SEM, could help discern the origin. Often, during slow crack growth the initial flaw might form an elliptic or semi-circular shadowed ring around the origin which will help locate its position (Fig. 1a). The presence of a flaw on a plane other than the fracture plane could also be distinguishable by CLSM as a shady region (Fig. 1b). Other surface markings like hackles and Wallner lines (hollow arrows in Fig. 1c) could also help establish the location of origin but might require some understanding of the dynamic behavior of the crack.

The exact placement of the fracture origin in general was expected to remain to some degrees of arbitrary. Nonetheless, the use of appropriate techniques could help reduce the uncertainty about the fracture origin location. Further discussion on how to determine the fracture origins could be found in the NIST guidelines [3].

2.3 Mirror-mist boundary determination

In this section, a novel method to establish the mirror radius based on CLSM profile images of the fracture surface is introduced. It should be noted, that the use of profilometry data to estimate fractographical features is not new. For instance, Duckworth et al. [6] used a Talysurf surface profilometer to trace the fracture surface along a direction parallel to the tension surface. The trace was then used to define the mirror boundary based on relative surface roughness. More recently, López-Cepero et al. [11] applied CLSM to study fracture profiles of erbia-stabilized zirconia and ruby fibers. Nonetheless, both these works established the length of the mirror region based on a threshold near the free surface rather than using a larger portion of information available from the fracture surface.

As the transition from mirror to mist region is a gradual progress caused by the gradual increase in surface roughness, an arbitrary roughness threshold had to be set to uniquely establish an arbitrary yet reputably measurable "mirror-mist boundary". In this work, the onset of the mist was chosen in the range 0.4-0.8 μm (i.e. the wavelength of visible light) based on the arguments similar to Quinn [4].

The surface contour at the stress-free surface is shown in Fig. 2, and basic fractographic feature could be inferred by the degree of roughness of the surface contour.

Detecting the mirror-mist boundary region from a fracture profile could be facilitated if the expected shape of the mirror-mist boundary could be described analytically. In particular, an analytical description of the mirror-mist boundary could enable the determination of the mirror radius using larger portions of the fracture surface rather than simply relying on limited information at the free-edge (which might also be affected by residual stresses). In this work the geometrical description of the mirror-mist boundary was based on fracture mechanics

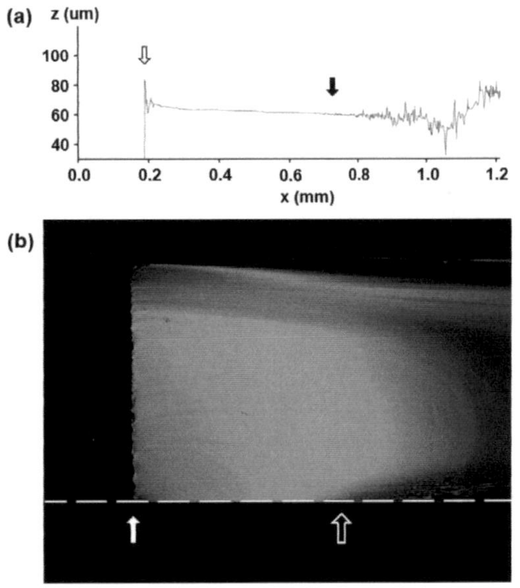

Figure 2. Profilometry contour (a) taken from the sample on the free edge (b, white dash line). White arrows denote the origin, black arrows denote the onset of mist.

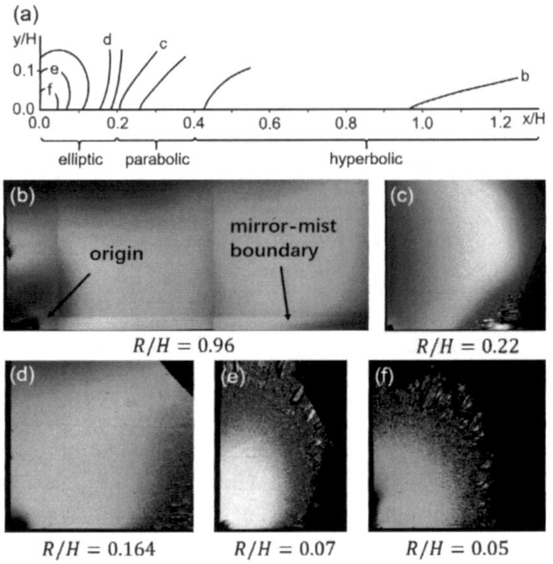

Figure 3. Mirror-mist boundaries under different fracture strength by 4PTB.

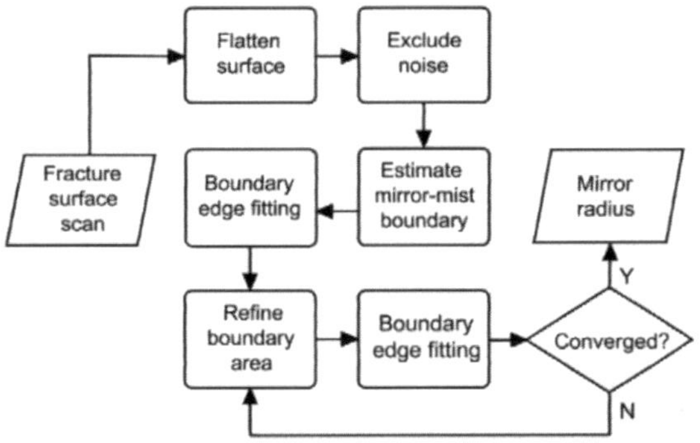

Figure 4. Flowchart for the algorithm used to determine the mirror radius.

analysis conducted by Dugnani and Zednik [12]. Dugnani and Zednik indicated that the geometric shape of the mirror-mist boundary for a brittle sample fractured in either tension or bending could be described by the equation of a conic section. In particular, the mirror-mist boundary shape near the free surface for samples fractured in bending was expected to follow an elliptical curve for values of $R/H < 0.2$, a parabolic curve for $0.2 < R/H < 0.4$, and a hyperbolic curve for values of $R/H > 0.4$. The shape of the mirror-mist boundary region for the soda-lime glass samples fractured in this work appeared to follow the trends described by the work of Dugnani and Zednik [12]. Fig. 3 (b) though (f) show examples of fracture surfaces for values of R/H corresponding to the elliptic, the parabolic, and the hyperbolic range. Also Fig. 3(a) shows a contour of the corresponding expected mirror-mist boundary shape for various values of R/H based on the theoretical work of Dugnai and Zednik [12]. With these premises, an efficient computational algorithm to obtain the mirror-mist boundary could be developed.

Fig. 4 shows the flowchart describing the methodology to assess the mirror radius. First, the surface roughness data in an area including the origin and the mirror-mist boundary were obtained by CLSM. Next both the estimated mirror-mist boundary radius, R_e (e.g., white dash line in Fig. 5b), and the slope of the mirror-mist boundary at the free surface, θ_m, were estimated visually. Alternatively, for $R_e > 0.16H$, the approximate slope of the mirror-mist angle, θ_m, could be approximated using the relationship suggested by Dugnani and Zednik [13]:

$$\frac{H}{R_e} = 2\tan(1.1\theta_m) + 0.25. \qquad (2)$$

After estimating the mirror radius, a rectangular region of height $H/10$ and length $2/5R_e$ was set at a distance of $0.05H$ from the origin and the free surface was selected to flatten the fracture surface and establish a reference plane (Fig. 5a). Next, all portions of the image that displayed a height difference greater than 30 μm from the reference plane were excluded. A parallelogram (black in Fig. 5b) with one edge centred at R_e, and height $0.1H$, width $0.5R_e$, and slanted edges at an angle θ_m from the free surface was then defined. The root-mean-square (RMS) of the surface height was calculated within the parallelogram and all the locations corresponding to surface roughness in range of 0.4-0.8μm were selected (white dots in Fig. 5c or black in Fig. 5d) and the remaining area discarded. A conic curve (black, Fig. 5c) was regressed through the points corresponding to the surface roughness in the range of interest. The type of conic section equation used to regress the data was determined based on the value of R_e/H, as explained in Dugnani and Zednik [12]. The curve obtained from the initial data regression (white, Fig. 5e) was used to define a new parallelogram with two curved edges and centred at the intersection between the fitted conic equation and the free surface (black, Fig. 5e). Subsequently, a new conic curve was generated, a new value of R_i obtained

https://doi.org/10.1051/matecconf/201816601005

and a new parallelogram defined. These steps were repeated until convergence which was achieved when the absolute value of the average difference between two subsequent iterations of the fitted conic curves was less than 1% (converged white curve in Fig. 5f).

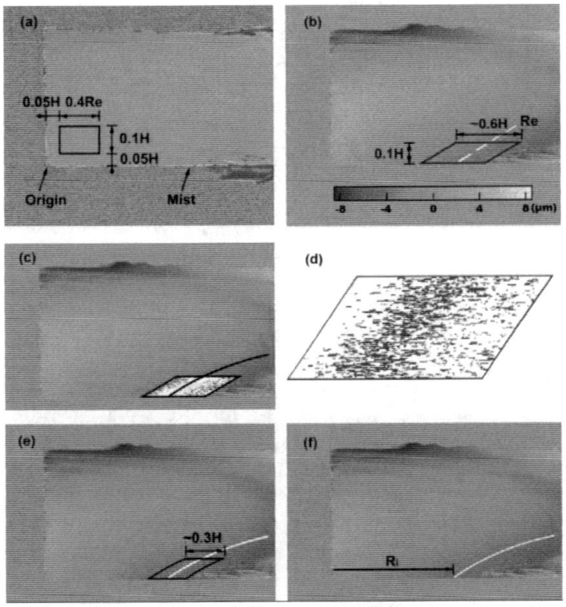

Figure 5. Main steps for algorithmic method.

3 Results

The mirror-mist boundaries were established using the profilometry-based measurements on twelve specimens using bright-field illumination and on six specimens using DIC. The constants A and C in Eq. (1) were obtained separately for both the proposed method and the traditional microscopic method using bright-field illumination measurements as suggested by the ASTM standard [2]. Additionally, mirror radii from the same specimens were also estimated by eight individual observers using standard optical microscopic technique and the profilometry-based method was repeated around eight times for each sample. These results were used to compare the repeatability of the measurements.

A relative stress estimation errors (RSE) were defined as:

$$\mathrm{RSE} = \frac{(\sigma_{estimated} - \sigma_{applied})}{\sigma_{applied}} \cdot 100\%. \qquad (3)$$

The value of RSE was computed for both the newly developed method and for the standard optical method of defining the mirror radii. In Eq. (3), $\sigma_{applied}$ stood for the recorded fracture strength from 4PTB, and $\sigma_{estimated}$ for the strength estimate by the measured mirror radius using Eq. (1). RSE represented the relative estimating error of the strength.

Table 2 shows different RSE from the profilometry scans and for the same samples based on optical images taken under the same viewing conditions. For the same set of samples, the deviations of RSE by optical microscope were larger than 3%, while for the profilometry-based algorithm deviations were less than 2%. Furthermore, DIC displayed a difference in mean RSE value up to 1.75% for optical method but it was nearly unchanged for the profilometry-based method.

Table 2. Selected RSE results for profilometry-based method vs. RSE from optical microscopy by eight observers.

Viewing modes	Samples tested	RSE (optical microscopy)	RSE (profilometry-based method)
Bright-field 10×	12	0.00%±3.14%	0.00%±1.84%
DIC 10×	4	1.75%±3.51%	0.11%±1.78%
DIC 20×	2	0.72%±3.16%	0.05%±1.86%

4 Discussion

Profilometry measurements carried out on twelve samples were compared with the corresponding optical methods for two viewing methods (i.e., bright field and DIC) and two magnification factors (i.e., 10x and 20x). Eight individual observers estimated the mirror radius per ASTM C1678 by optical microscopy (table 2).

Based on the analysis performed, the profilometry-based method was more repeatable than the optical methods.

In optical methods, observers gave different results due to their own interpretation of the location of the mirror-mist boundary, and the standard deviations in the reported strength for the data set considered were more than 3%. The reported results were also affected by the viewing modes and magnifications. For the profilometry-based method, the RSE deviations were within 2%. Furthermore, the mean value of RSE was higher for the DIC compared to bright-filed imaging, whereas it was nearly unchanged for the profilometry-based method.

No low-strength samples were examined in this manuscript, but the merit of this method could be more obvious in analyzing low-strength samples, since relative errors introduced by common viewing modes would be magnified on the low-strength basis. The mist could be faint near the free surface for low-strength samples, thus would lead to a large uncertainty in methods based on optical estimations. Nonetheless, the fitting process in the profilometry-based method could amend the boundary based on roughness and the clearer part of the boundary along with the curve shape indicated by Dugnani and Zednik.

The provided algorithmic method could have a small deviation (less than 2%) when defining the mirror-mist boundary, but the difficulty existed in dealing with very thin and flat mist, since the less points of the area of interest analysed, the more uncertainty existed in the final fitting result. Subjectivity also remained in the determination of mirror radius due to the complexity in defining the origin. The profilometry-based method in this manuscript introduced relatively stable and repeatable way to determine the mirror radius and to estimate the strength compared to conventional microscopy method.

5 Future Work

In this manuscript, an algorithmic method for mirror radius determination was provided based on the profilometry data collected by CLSM. Although the method was capable of identifying the mirror-mist boundary, and measuring the mirror radius in a more repeatable way compared to optical estimations, future work is recommended. In particular, future work should include:

(1) The fracture origin was determined with some degree of subjectivity. Further work should be carried out to estimate the origin more objectively based on the surface marking on the fracture surface (e.g., Wallner lines).

(2) One of the premises of this method was the shape of the mirror-mist boundaries followed a conic relation. This work conducted by Dugnani and Zednik [12] was based on the straight crack solution for the crack propagation. A more appropriate dynamic model should be implemented to better explain the loci of the mirror-mist boundaries.

6 Conclusion

In this manuscript, an algorithmic method was introduced to determine the mirror-mist boundary based on profilometry data obtained from the confocal laser scanning microscope. Fracture surface scans were obtained by confocal laser scanning microscope for the analysis of surface height and roughness. The mirror-mist boundary was determined based on the threshold of the roughness transition on the fracture surface. Comparison was made on twelve samples applying both profilometry-based method and optical microscopy conducted by eight individual observers per the ASTM C1678. Compared to the optical microscopy methods, the profilometry-based method displayed approximately 2% repeatability for the inner mirror measurements whereas conventional methods showed more than 3%. More importantly, the estimate from profilometry-based method was not affected by either viewing mode or magnification unlike the optical microscopy method. The profilometry-based analysis has then been shown to significantly reduce errors commonly introduced by the subjective interpretation of the mirror-mist boundary position, viewing modes, and magnification factors.

7 Acknowledgement

The authors would like to extend their gratitude to Ying Wang, Xiangxing Lv, Zhengdong Shan, Weizi Li, Huan Sun from UM-SJTU JI, and Anthony Moulins from ETS in Montréal for helping collecting data. They also thank Anthony Moulins for generating some of the profilometry data.

References

1. L. Orr. Mater. Res. Stand. **12** (1972)
2. ASTM C1678-10, Standard Practice for Fractographic Analysis of Fracture Mirror Sizes in Ceramics and Glasses (2015)
3. G.D. Quinn. Fractography of Glasses and Ceramics V: Ceramic Transactions. **199** 163-187 (2007)
4. G.D. Quinn: *NIST Recommended Practice Guide: Fractography of Ceramics and Glasses*. NIST SP - 960-16 (2016)
5. J. Mecholsky, S. Freiman: *Determination of fracture mechanics parameters through fractographic analysis of ceramics*, in *Fracture Mechanics Applied to Brittle Materials*: ASTM International, 136-150 (1979)
6. W. Duckworth, D. Shetty, A. Rosenfield. Glass Technol. **24** 263-273 (1983)
7. D. Hull. J. Mater. Sci. **31** 1829-1841 (1996)
8. D. Hull: *Fractography: observing, measuring and interpreting fracture surface topography*. Cambridge University Press (1999)
9. D. Kulawansa, L. Jensen, S. Langford, J. Dickinson, Y. Watanabe. J. Mater. Res. **9** 476-485 (1994)
10. ASTM C158-02, Standard Test Methods for Strength of Glass by Flexure (Determination of Modulus of Rupture) (2012)
11. J. Lopez-Cepero, A. De Arellano-Lopez, J. Quispe-Cancapa, J. Martínez-Fernández. **19** 13-15 (2005)
12. R. Dugnani, R.J. Zednik. Eng. Fract. Mech. **165** 87-97 (2016)
13. R. Dugnani, R.J. Zednik. **97** 3853–3856 (2014)

Flexible Hinge Parameter Design and Performance Simulation of Vibration Assisted Machining Platform

Yixiong Yan[1]

[1]*Central South University, Mechanical and Electrical Engineering Department,410083Changsha Hunan, China*

Abstract. Vibration assist machining is a modern mechanical manufacturing technology which is aimed at achieving an efficient high-quality machining. This article focuses on a new-type vibration assisted machining platform which based on the stiffness and frequency requirement to obtain the key structural parameter and nature frequency of the platform by using the finite element method. Subsequently, static strength of the platform has been analyzed. Moreover, different excitation frequencies of the piezoelectric ceramics have been loaded at two vertical directions. The trajectory stability and characteristics of the platform under different actuating frequency have been figured out by using the finite element multi-body contact transient response analysis method. The results show that the nature frequency excitation results in large and destructive displacement. The steady-state response trajectory of the platform is elliptical under different excitation frequencies while under high excitation frequency it would become an elliptical moving. Once the unstable vibration time of platform increase, the unstable vibration amplitude would increase. The research results can provide a basis for the development of a new-type vibration assisted machining platform and the processing technology.

1 Introduction

Vibration assisted machining has become one of the most important technologies for precision and ultra-precision machining. During processing, it is advantageous to have controlled vibration with high frequency and low amplitude [1]. Surface and subsurface damage would take place due to the brittle fracture during the traditional process [2]. The vibration assisted machining can reduce machining force in the process, resulting in an increasing of machining surface accuracy and reduction of tool wear [3]. In recent years, vibration assisted machining technology has extensive applications in ultrasonic vibration cutting, grinding, plastic processing with its rapid development. This technology also solves a considerable number of problems in relation to difficult-to-process materials machining. It can be predicted that the application would be more and more widely [4].

The vibration assisted machining platform is the key component to achieve vibration assisted cutting. In order to meet the requirements of high efficiency and high-quality cutting, this paper designs the key structure of the flexible hinges which functions as the transmission component of the platform. Moreover, the dynamic and static performance of the platform have been analysed by using the finite element analysis simulation. It lay the foundation for the development of a new type of vibration assisted platform.

2 Structure and components of the vibration assisted platform

Vibration assisted machining platform is used for cutting, grinding and other processing to improve the processing technology [5]. The platform mainly consists of piezoelectric ceramic placement slots, flexible hinges and a stage. The structure is shown in Figure 1.

In order to ensure that the flexible hinges work without cross-coupling displacement and the vibration can be smoothly conducted in both directions, the structure of the flexible hinges is designed as a double-layer structure. The inner layer hinges guide the vertical vibration while the outer layer hinges guide the horizontal vibration.

The material used for the platform is 30CrMnSiA alloy structural steel, its material performance parameters are shown in Table 1.

The actuation components are piezoelectric ceramics, mounted in two perpendicular directions [manufactured by PhysikInstrumente (PI) GmbH & Co.KG, Germany]. According to their model parameters and platform quality parameters, the stiffness of the platform and the limitation of natural frequency are obtained. Stiffness should be less than 30N/um, while the natural frequency should be greater than 2500Hz.

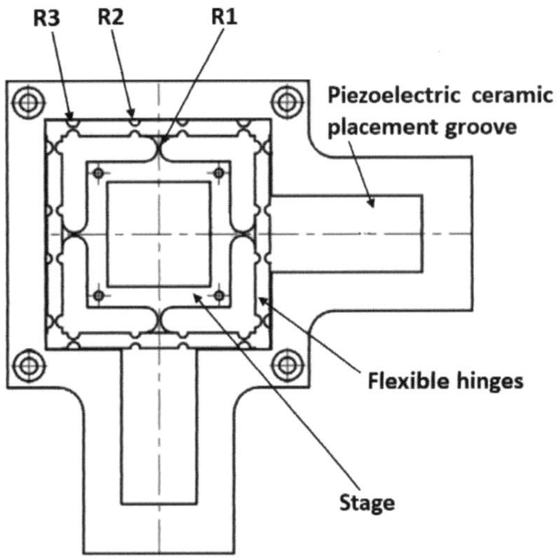

Figure 1. The structure of the platform

Table 1. Carbon structural steel performance parameters

Property	Value	Unit
Density	7850	kg/m^3
Young's Modulus	1.96E+11	Pa
Poisson's Ratio	0.3	-
Bulk Modulus	1.6333E+11	Pa
Shear Modulus	7.5385E+10	Pa

3 Structural parameters of flexible hinges

3.1 Flexible hinges components structure analysis

Figure 2 shows the structure of the flexible hinge, where b is the width of the flexible hinge, t is the minimum thickness of the middle part of it, R is the radius of the notch, L is the length of it, the rotational stiffness on the z-axis and the tensile stiffness on the x-axis is $k_{\alpha_z}$and k_x, respectively. The minimum thickness t and the width b are fixed values, the angular deformation △x can be expressed by the stiffness equation (1) [6].

$$k_s = \frac{F_x}{\Delta x} = \frac{Eb}{\left[\frac{2(2s+1)}{4s+1}\arctan\sqrt{4s+1} - \frac{\pi}{2}\right]} \quad (1)$$

Where s=R/t. It can be seen that the stiffness is mainly affected by the cutting radius R. And the flexible hinge component is symmetrical as shown in Figure 1, due to the fact that flexible hinge component is symmetrical, only three notch radii R1, R2 and R3 require to be considered which are marked in figure 1.

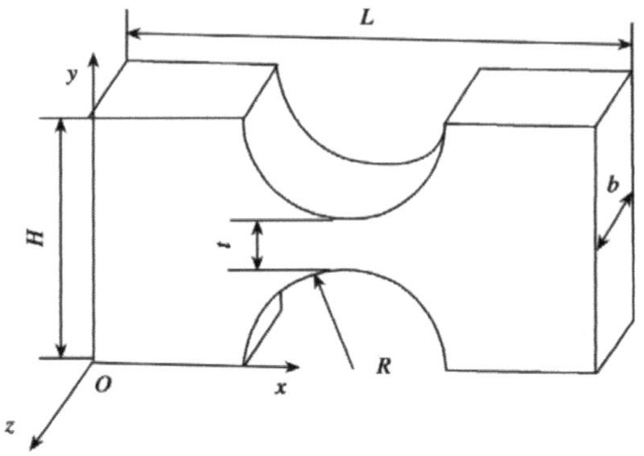

Figure 2. Structure of the flexible hinge

3.2 Flexible hinge parameter design

In order to figure out these three notch radii, several works should be done. First, a clear and definite goal should be defined as it is the basis, followed by the building of a mathematical model. With this model, ANSYS Workbench is used to work out the parameters. The greater stiffness of the platform means that the piezoelectric ceramics need to provide greater driving force, which will increase the operating temperature. As a result of high operating temperature, the actuator performance will be degraded. Therefore, the stiffness of the platform should be as small as possible [7]. In order to widen the working frequency range of the platform and better avoid the resonance effect, the natural frequency of the platform should be as high as possible [8]. At the same time, the relationship between the stiffness and the natural frequency of the platform are expressed by equation (2).

$$\omega_0 = \frac{1}{2\pi}\sqrt{\frac{k_x}{M}} \quad (2)$$

It can be seen that when the mass of the platform is fixed, the value of the natural frequency and stiffness is proportional. In summary, the objective of the solution is to ensure the natural frequency of the platform is as large as possible and the stiffness is as small as possible under the condition of mutually restrictive.

The range of R1, R2, R3 according to the overall structure design decision are shown as equation (3).

$$\begin{cases} 2 < R1 < 5 \\ 0.5 < R2 < 2.9 \\ 0.5 < R3 < 2.9 \end{cases} \quad (3)$$

The mathematical model can be expressed as equation (4):

$$\begin{cases} Max\, f = f(R1, R2, R3) \\ Min\, k = k(R1, R2, R3) \end{cases} \quad (4)$$

After the mathematical model is built, the established platform model is imported into the ANSYS Workbench. The range of the notch radii is set as equation (3). Then the software generates random and eligible 2000 groups R1, R2, R3 to obtain the corresponding stiffness and natural frequency of the platform. The optimal result is presented in Table 2.

Table 2. Parameter solving result

R1	R2	R3	Stiffness	Natural frequency
4.95mm	1.14mm	1.71mm	27.7N/um	2 625Hz

4 Static performance of the platform

The flexible hinges of the platform will be subjected to the dynamic force of the actuator during operation, and its strength must be able to withstand the peak thrust of the actuator.

The designed platform model is imported into ANSYS Workbench. Hexahedron-dominated method is selected as meshing method while approximate and curvature control method in the advanced dimension control set group is used to mesh the curve structure. The minimum division size is 0.16mm, and a total of 219493 units are obtained. The maximum working thrust is 2000N, which is applied to the platform, and the static analysis result is shown in Figure 3.

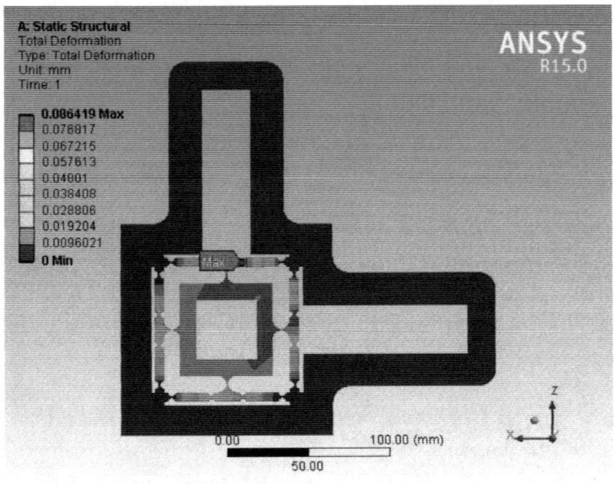

a) Deformation distribution

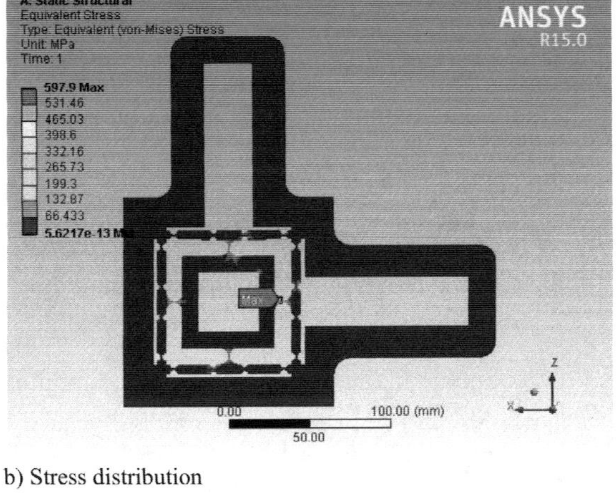

b) Stress distribution

Figure 3. Static performance of the platform

It can be seen from the simulation result that the maximum deformation of the platform is 0.086mm. The main deformation occurs at the flexible hinges, and the deformation is relatively reasonable. The maximum stress value is 597.9Mpa while the yield point of the material is $\sigma_s = 1080Mpa$. Thus the safety factor is greater than 1.5, so that the static performance of the platform meets the requirement.

5 Transient response analysis of the platform

5.1 Transient response analysis

The transient dynamics solution uses the modal superposition method to derive eigenvectors. By multiplying and adding the eigenvectors. The response of the structure can be solved [9]. Through the transient response analysis, the vibration rule of the platform under different external excitation can be obtained.

The platform model is imported into the ANSYS Workbench. Two vertical external excitations, generated by piezoelectric ceramics, are applied to the platform based on the model analysis. The mechanical equation of the excitations are expressed as follow.

$$F = A\sin(2\pi f t + \varphi) \tag{5}$$

$$F = A\cos(2\pi f t + \varphi) \tag{6}$$

Considering that the natural frequency of the platform is 2625Hz, four groups of excitation frequencies are selected for simulation: 500Hz, 2000Hz, 2625Hz, 3000Hz. In accordance with the requirements of the platform, the amplitude value is set as 4um. Four groups of results are shown as below.

5.1.1 The transient response analysis result under 500Hz excitation

The transient response in the X-axis direction of the reference point on the stage of the platform when the excitation frequency is 500Hz is shown in Figure 4. The platform keeps unsteady vibration within 0.1 second while the maximum amplitude reaches 7.3329E-03mm. Moreover, the vibration began to stabilize when near 0.2 second. The deformation response trend in Z-axis direction is very similar to the X-axis.

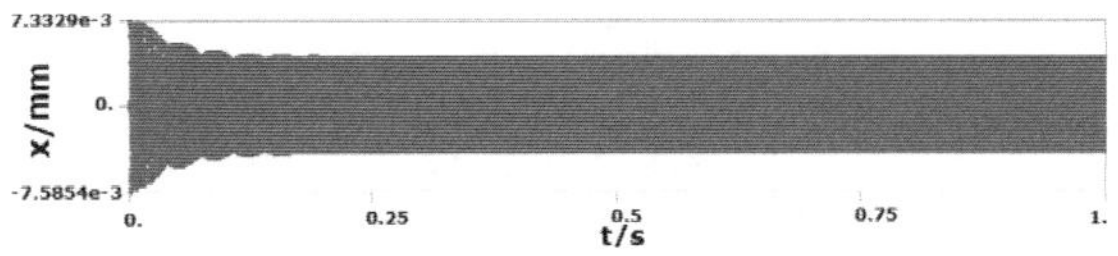

Figure 4. The deformation response in X-axis direction when loading 500Hz excitation

 https://doi.org/10.1051/matecconf/201816601006

According to the displacement data in X and Z direction, the trajectory chart of the reference point on the platform can be obtained in Figure 5. The 'wave' line refers to the trajectory in the unstable period, and the 'stable' line refers to the trajectory in the stable period. It can be seen that the trajectory in unstable period is similar to a rectangle, which transpires at the both ends of the X-axis. While the trajectory in the stable period basically coincide as a smooth ellipse, and the long axis of the ellipse is at an about 45°to the X-axis.

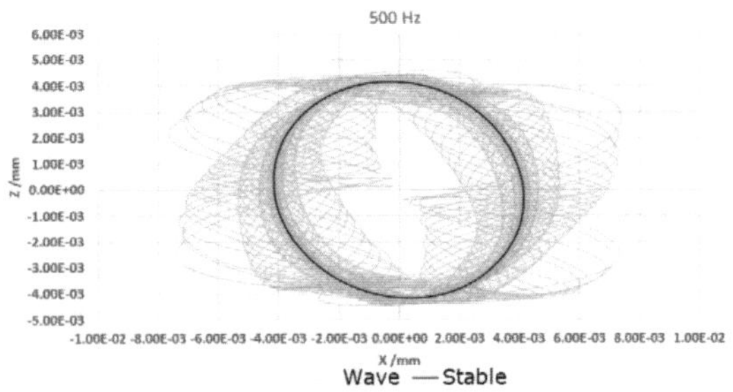

Figure 5. Trajectory chart of the deformation during 500 Hz

5.1.2 The transient response analysis result under 2000Hz excitation

The transient response in the X-axis direction of the reference point on the stage of the platform when the excitation frequency is 2000Hz is shown in Figure 6. The maximum amplitude of the unstable period is 2.009E-02mm. The amplitude is greater than before, and the unstable period lasts longer. However, the transition from unstable to stable vibration is quite smooth.

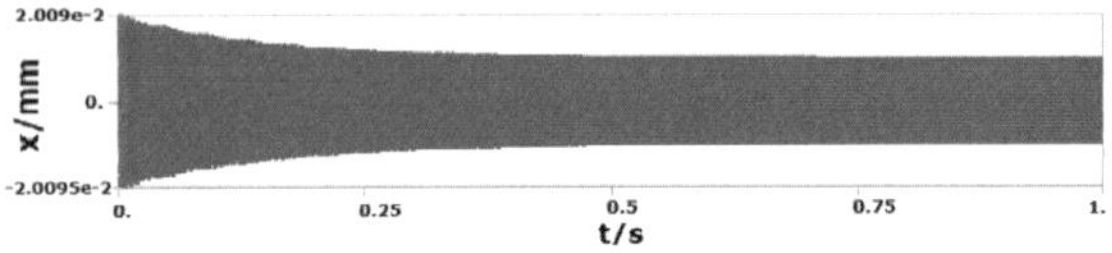

Figure 6. The deformation response in X-axis direction when loading 2000Hz excitation

The reference point displacement in X-Z plane is shown in Figure 7. The unstable period trajectory is similar to an ellipse. However, the route trend is not obvious. The overlap rate of the unstable period trajectory is higher than before. The stable period trajectory coincide as a smooth ellipse while the axes of it are parallel to the X-axis and Z-axis.

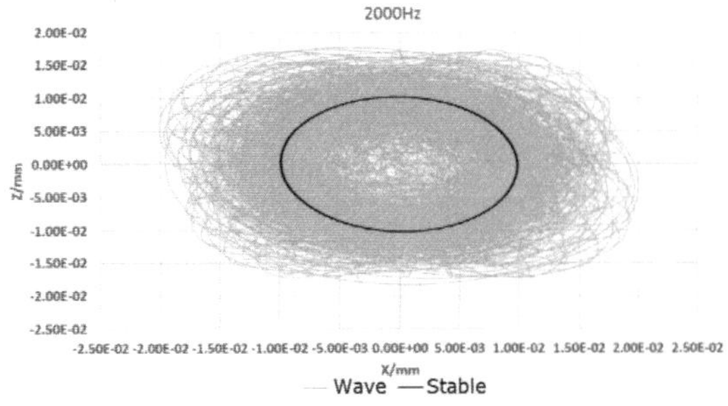

Figure 7 Trajectory chart of the deformation during 2000Hz

5.1.3 The transient response analysis result under 2625Hz (natural frequency) excitation

The transient response in the X-axis direction of the reference point on the stage of the platform when the excitation frequency is 2625Hz is shown in Figure 8. The unstable period increased significantly, and the maximum amplitude reaches a value of 1.0061mm. The strong vibration may cause damage to the platform so that it is essential to avoid such circumstance.

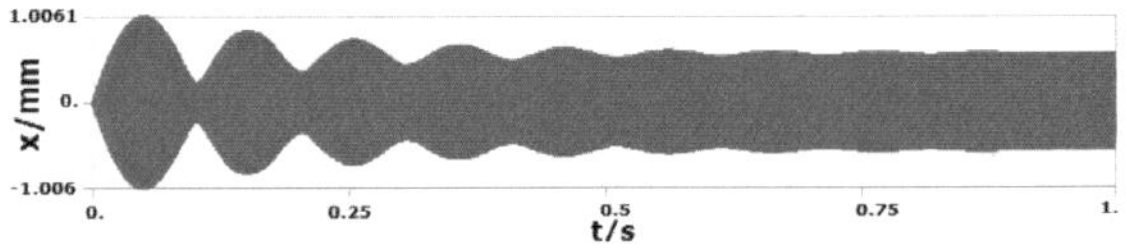

Figure 8. The deformation response in X-axis direction when loading 2625Hz excitation

The reference point displacement in X-Z plane is shown in Figure 9. The route trend is quite clear even in unstable period.

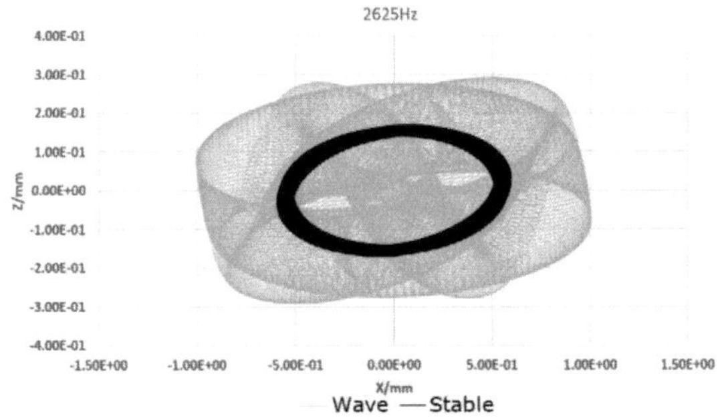

Figure 9. Trajectory chart of the deformation during 2625Hz

5.1.4 The transient response analysis result under 3000Hz excitation

The transient response in the X-axis direction of the reference point on the stage of the platform when the excitation frequency is 3000Hz is shown in Figure 10. The unstable period is much shorter than last one, and the maximum amplitude of the unstable period is 2.7283E-02mm.

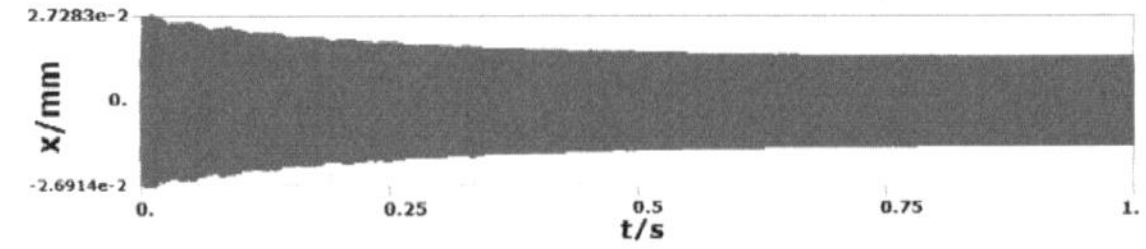

Figure 10 The deformation response in X-axis direction when loading 3000Hz excitation

The reference point displacement in X-Z plane is shown as Figure 11. It can be seen that the unstable period route trend is unclear, but the unstable period is shorter than last one. The trajectory in stable period still coincide as a smooth ellipse.

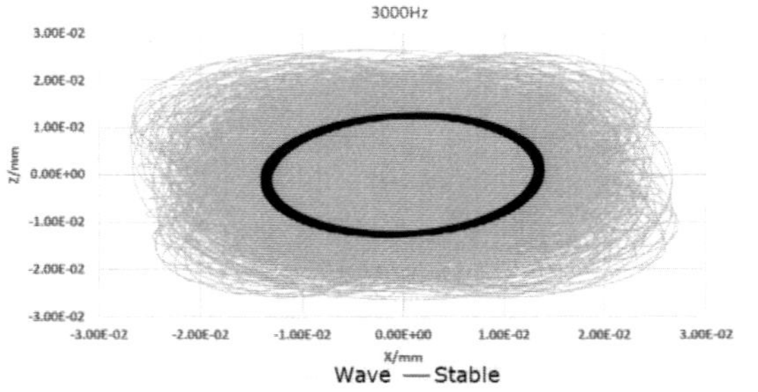

Figure 11. Trajectory chart of the deformation during 3000Hz

The response characteristics of the platform in different operating frequency can be conclude in the table below.

Table 3. Response characteristics at different excitation frequencies

Frequency /Hz	Duration in unstable period /s	Maximum amplitude in unstable period /mm	Amplitude in stable period /mm
500	0.208	7.3E-03	4.3E-03
2 000	0.625	2.0E-02	1.2E-02
2 625	0.875	1.006	0.759
3 000	0.812	2.7E-02	1.3E-02

6 Conclusion

1) With the aim of small stiffness and high natural frequency, the key parameters of the flexible hinges structure is obtained by the finite element method. 2) The static strength and deformation of the platform under the external excitation produced by piezoelectric ceramics are calculate and the reliability of the static strength of the platform is ensured. 3) The finite element method is used to analyze the transient response of the platform under different frequency excitations. Moreover, the response characteristics of the platform has been presented: The trajectory of the reference point on the platform in the stable period coincide as a smooth ellipse when two excitations loading in two vertical directions. Under the excitation of natural frequency, the duration of the unstable period is long, and the amplitude of the vibration is harmful to the platform. It provides a basis for choosing different excitation parameters for different processing technologies.

Acknowledgements

The authors express their deep gratitude to the National Natural Science Foundation of China (NSFC) (No. 51575533)

References

1. W. Xu and L. Zhang, AIM, **3**, 173-192 (2015)
2. D.E. Brehl* and T. Dow, PE, **32**, 153-172 (2008)
3. T. Tawakoli, B. Azarhoushang and M. Rabiey, TIJOAMT, **42(9/10)**, 883-891 (2008)
4. Z. Zhou, Q. Zhou and Y. Ren, JOME, **46(17)**, 105 (2010)
5. X. Zhang and F. Jiao, TE, **1**, 3-8 (2012)
6. Z. Yu, X. Yao and X. Song, CJOSI, **30(9)**, 1818-1822 (2009)
7. B. Jaffe, W. Cook and H. Jaffe, JOSV, **20(4)**, 271-279 (1971)
8. H. Ji, and Y. Wen, AMR, **179-180**, 1368-1373 (2011)
9. Q. Zhang, X. Chen, Z. Zhu, et al., **85(3)**, 1819 (2013)

A Comparison of Effective Tension Calculation for Design Belt Conveyor between CEMA and DIN Standard

Iman Satria[1], Meifal Rusli[2]

[1,] *Mechanical Department Engineering, Faculty of Industry Technology, University of Bung Hatta Padang-Indonesia,*
[2] *Mechanical Department Engineering, Faculty of Engineering, University of Andalas Padang-Indonesia,*

Abstract. In the present era, many industries are demanding material transfer equipment that works maximally and cheaply. Belt Conveyor is one of the most efficient material transfer equipment compared to heavy equipment or other transportation equipment, because it can transport the material in large capacity. When designing conveyor belts, many industries want large transport capacity at low cost. One of the cost savings can be done with low motor power consumption. Many standards describe the effective tension calculation on the conveyor belt in detail with the parameters, since the effective tension calculation results are essential for determining motor power. This paper aims to analyze the effective tension on conveyor belts using CEMA 5th, CEMA 6th and DIN 22101 standards with case studies of 1000 TPH carrying capacity and 3 m/s conveyor speed. The difference between CEMA and DIN when determining the effective tension is on the use of coefficient of friction, because DIN uses a global coefficient of friction while CEMA uses different friction coefficients on idler, belt and material. This difference in coefficient of friction results in different ways of calculating the resistance being the accumulation of the effective tension. The result is an effective tension value using DIN 22101 standard of 22,148.16 N with motor power consumption of 66.5 kW, while effective tension value using CEMA 5th standard is 32.201.66 N with motor power consumption 96.47 kW, and effective tension value using CEMA 6th standard is 29,686.48 N with 89.06 kW motor power consumption.

1 Introduction

Material transport equipment is very important in this modern era of life, especially in helping a job to obtain maximum results. Material transport equipment serves to move material in certain areas, to a department, plant and plant, construction site, storage and loading.

Today, they are in most cases the most cost-effective solution for handling bulk material mass flow over short and medium conveying distances. Despite the already advantageous costs for belt conveyor operation, there is still a desire to reduce these costs even futher [9]. One of the cost savings can be done with low motor power consumption, where the motor power value is directly proportional to the effective tension value that occurs in a conveyor system.

Effective tension is defined as the sum of all the resistance of motion referring to the motor [6]. The resistances that affect the effective tension are friction resistance due to friction between the belt, idler and material when the conveyor belt is running. In addition, the energy resistance or resistance caused by the load carried by the conveyor itself is a resistance that also refers to the motor.

In determining the effective tension of the conveyor belt, there are many standards that can be noticed by engineer such as; CEMA, DIN 22101, and so forth. Standard is a norm which is usually a formal document that creates uniform engineering or technical criteria, methods, processes and practices. The two most common design methods used to calculate the effective tension on the conveyor belt are the Conveyor Equipment Manufacturer's Association (CEMA) method [1] [2] and the Deutsches Institute Normung (DIN) method DIN 22101 Standard [3].

Based on these problems, it is necessary to conduct research that aims to analyze the effective tension on the conveyor belt using standard variations DIN 22101, CEMA 5th, and CEMA 6th. Problems to be investigated in this research include: effective tension calculation analysis on belt conveyor capacity 1000 TPH using standard DIN 22101, CEMA 5th, and CEMA 6th, and determine difference of three standard in determining effective tension.

2 Belt conveyor data

In this research, an effective tension calculation analysis using CEMA standard and DIN 22101 standard with case study on conveyor capacity of 1000 ton / hour. Study of case in this research using data conveyor at PT. Pelabuhan Indonesia (Pelindo) II Bengkulu as above in table 1. Using the data obtained we will analyze the differences between the DIN 22101, CEMA 5th, and CEMA 6th standard in designing and analyzing the effective tension to determine the motor power of the conveyor system.

Table 1. Conveyor industrial data

Parameter	Value (S.I Units)	Value (British Units)
Length along belt	154 m	505 ft
Lift	17 m	56 ft
Angle of Incline	6.30°	6.30°
Belt Width	1.2 m	48 in
Skirt width	0.8 m	31.48 in
Capacity	1000 tph	1000 tph
Belt Speed	2.5 m/s	492 ft
Bulk Density	0.96 ton/m^3	60 lbf/ft^3
Carry idler spacing	1.2 m	4 ft
Drive pulley wrap angle	210°	210°
Coefficient friction (*Pulley/ belt*)	0.35	0.35
Belt Thickness	0.0144 m	0.567 in
No. of belt cleaners	1	1
length of skirtboard	6.456 m	21.18 ft
Idler diameter	139 mm	5.5 in
No. of roll per idler set	3	3
Troughing angle	35°	35°
Material Surcharge Angle	20°	20°
Angle of Repose	35°	35°
Idler Misalignment	0	0
No. of drive pulley	1	1
Diameter of drive pulley	660 mm	26 in
Shaft Diameter of Drive Pulley	136 mm	5.35 in
Weight of Drive Pulley	16000 N	3596.94 lbf
No. of Snub/Bend pulley	5	5
Diameter Snub/Bend pulley	420 mm	15.53 in
Shaft Diameter Snub/Bend pulley	80 mm	3.149 in
Weight of Snub/Bend pulley	8000 N	1798.47 lbf
Take-up tension	40754.95 N	9162 lbf
Type of belt used	*Fabric Belt*1200-EP 630/ 4ply 6+3	
Type of idler bearing used	*Ball bearing*	
Bulk Material	*Coal (Antrhacite Sized)*	

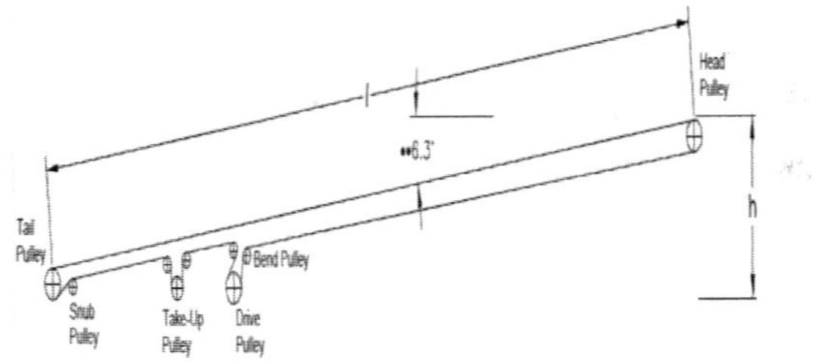

Figure 1. Scheme of conveyor

To analyze the effective tension calculation performed using computer program that is MATLAB 2008a.

3 Result and discuss

3.1 Comparison of Coefficient for DIN 22101, CEMA 5th and CEMA 6th Standard

The fundamental difference between DIN 22101 and CEMA 5th is on the use of friction factor, where the DIN 22101 standard uses a single or global friction factor (f) while the CEMA 5th standard distinguishes the coefficient of friction on idlers, belts, and materials (Kx and Ky). While CEMA 6th further refine the friction factor of CEMA 5th, where K_x is changed to K_{is}, K_{iV} and C_{iW}. The coefficient of friction is useful for calculating resistance due to friction that occurs in the belt, idler and material.

Table 2. Comparison of Coefficient for DIN 22101, CEMA 5th, and CEMA 6th methods.

Variable	DIN 22101	CEMA 5th	CEMA 6th
Friction Factor	f	K_x	K_{Iv}
			K_{is}
			C_{iW}
		K_y	T_{bi}, T_{im}, T_{mz}
Factor Correction Temperature	ct	K_t	K_{iT}
Primary Tension	F_H	T_x	T_{is}
			T_{iW}
		T_{yc}	T_{bi}
		T_{yr}	T_{im}
		T_{ym}	T_{mz}
Secondary Tension	F_N	T_p	T_p
		T_{ac}	T_{ac}
Secondary Tension Factor	C	-	-
Gradient Tension	F_{St}	T_m	T_H
		T_{am}	T_{am}

For DIN 22101 methods,

ct = Temperature correction factor, where (ct = 1 at 20°C, ct = 1.7 at T = -40°C)
f = Friction factor, value 0.01 < f < 0.04, based on operation and installation conditions.
C = Secondary tension factor, based on conveyor length (C = 1.92 at 80 m, C = 1.05 at > 2 km)

For CEMA 5th methods,

Kx = Idler friction factor (See the CEMA equation)
Ky = Belt and material flexure factor (0.016 to 0.035), from CEMA table.
Kt = Temperature correction factor, where Kt = 1 (0 < T < 40°C), Kt = 3 (T = -40°C)

For CEMA 6th methods,

K_{is} = Seal torsional resistance per roll (3 < K_{is} < 7.25) based on the type of idler used.
K_{iV} = Torsional speed effect, (0.003 to 0.004), based on the type of idler used.
K_{iT} = Temperature correction factor (See the CEMA equation)
C_{iW} = Torsional load effect (0.00125 to 0.0029), based on the type of idler and bearing used.

3.2 Effective tension calculation result

Base on DIN 22101 methods, the result shown as figure above.

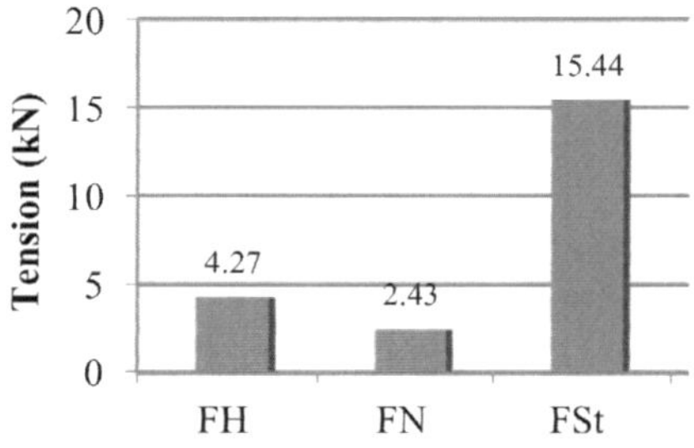

Figure 2. Graph of comparison of tension used DIN 22101 standard

The greatest tension occurs due to the energy load (elevation tension), where the value of the gradient tension (FSt) is 15441.2 N. While the secondary tension (FN) is the smallest tension that occurs with the value of 2433.92 N. The tension due to the primary tension (FH) occurs of 4273.04 N.

CEMA 5th methods,

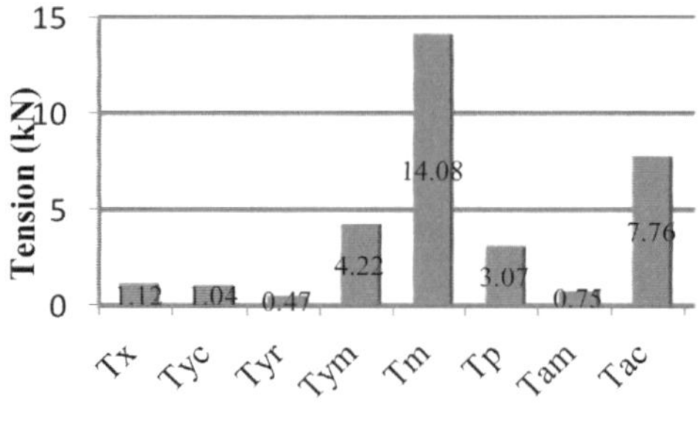

Figure 3. Graph of comparison of tension used CEMA 5th standard

Fig 3 shows the results of the tension calculations that occur using the CEMA 5th standard, where the greatest energy loss is caused by the tension lift or lower of material (Tm) with a value of 14078.98 N. Tx, Tyc, Tyr, and Tym are the tension caused by friction on the idler, belt and material occurs of 6854.26 N. The tension due to the accessories tension (Tac) occurs of 7763.24 N.

CEMA 6th methods,

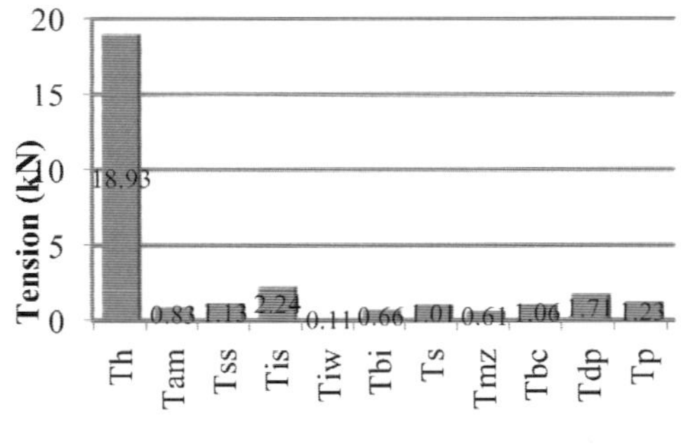

Figure 4. Graph of comparison of tension used CEMA 6th standard

Fig 4 shows the results of the tension calculations that occur using the CEMA 6th standard, where the greatest energy loss is caused by the tension lift or lower of material and belt (Th) with a value of 18926.8 N. Tis, Tiw, Tbi and Tmz are the tension caused by friction on the idler, belt, and materials. The value of tension occurs due to the primary friction on belt conveyor (Tis, Tiw, Tbi and Tmz) is 3629.5 N. While the accessories tension (Tss, Ts, Tbc, Tdp and Tp) is occurs of 6298.33 N.

3.3 Comparison of Effective Tension Result

Table 6 shows the comparison of effective tension calculations using DIN 22101, CEMA 5th, and CEMA 6th standards.

The effective tension value is directly proportional to the motor power consumption value, so that the motor power value using DIN 22101 standards is 66.5 kW, the motor power value using CEMA 5th standard is 97.42 kW and the motor power value using CEMA 6th standard is 89.06 kW. Seen from the three largest energy loss standards caused by the tension of energy (height) and then followed by the primary tension and the smallest is secondary tension.

Table 6. Comparison of effective tension with DIN 22101, CEMA 5, CEMA 6 standard

Variable	DIN 22101		CEMA 5		CEMA 6	
	Value	Symbols	Value	Symbols	Value	Symbols
Primary Tension	4.27 kN	FH	6.85 kN	(Tx+Tyr+ Tyc+Tym)	3.63 kN	(Tis+Tiw+ Tbi+Tmz)
Secondary Tension	2.43 kN	FN	10.83 kN	(Tac+Tp)	6.29 kN	(Tac+Tp)
Gradient Tension	15.44 kN	FSt	14.83 kN	(Tm+Tam)	19.76 kN	(Th+Tam)
Effective Tension	22.15 kN	Fw	32.5 kN	Te	29.69 kN	Te
Motor Power	66.5 kW	Pw	97.42 kW	P	89.06 kW	P

3.4 Result of maximum tension calculation with variation capacity using cema standard 6th

The maximum tension is the largest tension that occurs in a conveyor belt system. The purpose of calculate the maximum tension in order to determine the construction specifications according to the maximum load value that occurred. Here is a graph showing the maximum load value on a conveyors variation with a conveyor speed of 3 m/s.

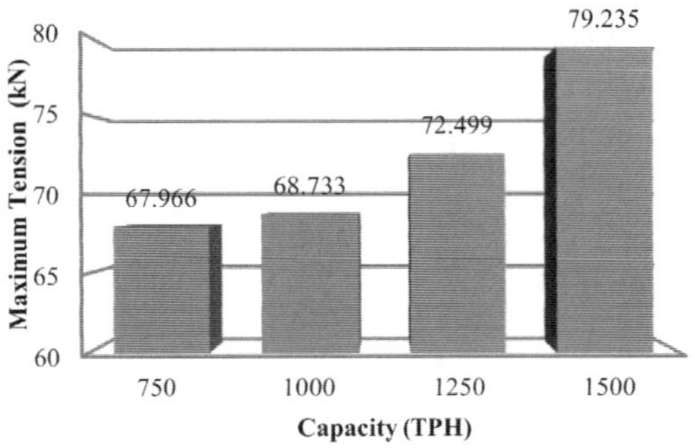

Figure 5. Graph of maximum tension as function capacity of belt conveyor

4 Conclusion

After doing research and data analysis of test result can be concluded that:

1. Effective tension value and lowest motor power consumption is calculation using DIN 22101 standard with effective tension value of 22.15 kN and motor power consumption of 66.5 kW. While the effective tension value and highest motor power consumption is an effective tension calculation using CEMA 5th standard with effective tension value of 32.2 kN and motor power consumption of 97.42 kW
2. The fundamental difference of the effective tension calculation of CEMA and DIN 22101 is located on the coefficient of friction, where DIN standard uses single friction coefficient (f) while CEMA standard differentiates coefficient of friction on belt, idler, and material (K_x and K_y). In CEMA 6th Kx developed again into K_{is}, K_{iV} and C_{iW} to find the voltage on the idler, and K_y replaced by calculating the tension equation of belt (T_{bi}, T_{im}, T_{mz}).
3. From the effective tension calculation using the three standards, the highest tension increase is caused by the tension due to the energy.
4. The result of the maximum tension analysis using CEMA 6 standard is directly proportional to the variation of the carrying capacity. Where the greater the carrying capacity it will increase the maximum tension value. The highest maximum tension value is 79.23 kN at the carrying capacity of 1500 TPH, while the lowest maximum tension value is 67.96 kN at 750 TPH carrying capacity.
5. In terms of safety calculation results using the CEMA 5 standard is more secure than the standard DIN 22101 and CEMA 6 because it has the highest motor power value with a value of 96.47 kW. However, in terms of economic value, the calculation result using DIN 22101 standard is cheaper compared to CEMA 5 and CEMA 6 standards because it has a much lower motor power value with a value of 66.5 kW.

References

1. Conveyor Equipment Manufacturers Association (CEMA) 2002. *Belt Conveyor for Bulk Materials. 5th Edition.*
2. Conveyor Equipment Manufacturers Association (CEMA) 2007. *Belt Conveyor for Bulk Materials. 6th Edition.*
3. Deutsches Institut fur Normung (DIN) 22101. 2002. *Belt Conveyor for Loose Bulk Materials.*
4. Gerrard, Bruce dan O'Rourke, L. 2009. *Optimisation of overland conveyor performance.* Australian Bulk Handling Review.
5. Gupta, A. 2014. *An investigation into tensile force of long distance belt conveyor.*
6. Horrison, A. 2009. *A Comparison of Friction Models for Conveyor Design*. Bulk Solid & Powder – Science Technology Vol. 4 No.1.
7. Kulinowski, P., Kasza, P. *Belt Conveyor Calculation*. Department of Mining, Dressing and Transport Machines AGH.
8. Kulinowski, P., Kasza, P. *Belt Conveyor for Bulk Materials Calculations by CEMA 5th Edition*. Department of Mining, Dressing and Transport Machines AGH.
9. Lauhoff, H. 2005. *Speed Control on Belt Conveyors – Does it Really Save Energy?*. Bulk Solid Handlings Vol. 25 No.6
10. Visser, T. H. A. 2015. *Energy Prediction Models of Belt Conveyor System Towards Speed Control.* Department Marine and Transport Technology.

Design and Experimental Study of Rotary Type High-Speed Shearer for Metal Bars

Renfeng Zhao [1], Jingyu Zhou [1], Yan Li [1], Shuqin Fan [2], Jingxiang Li [2], Xudong Xiao [1] and Zhi Yang [1]

[1] *Xi'an University of Technology, College of mechanical and precision instrument engineering, xi'an, 710048, China*
[2] *Xi'an Jiaotong University, School of Mechanical Engineering, xi'an, 710049, China*

Abstract: For the precision shear of bars, the kind of rotating high-speed precision cutting method has been given. The structure design of rotary precision shearing machine and the structure design of feeding structure are completed. The method of numerical simulation is used to optimize the cutting tool, and then the experiment platform is completed. Rotary impact and the stress concentration effect of the surface of the bar are used in the paper. The experiment is carried out for the Φ8 mm diameter stainless steel bar stock by changing the speed of the cutting tool through the converter and controlling rod feeding efficiency. The experimental results show that the method is feasible to meet the requirement of various material of the small diameter rod in the industry.

Keywords: Bar; High-speed Shearing; Stress Concentration; Cross-section Quality

1 Introduction

Metal bars are widely used in the industrial production. The traditional blanking methods which are sawing, turning and punching and shearing blanking[1-3]can not meet the new requirements of industrial production. Therefore, it is of great significance to study the new blanking method. But, the sawing and cutting blanking are not only a waste of raw materials, but also the efficiency is low. The quality of impact shear blanking method is poorer. High energy consumption blanking process, such as laser cutting can only cut the plate and pipe with medium and small thickness, blanking efficiency is low, and equipment cost is high. The high pressure water jet cutting equipment is expensive, the abrasive price is higher and lower precision than laser cutting. The method of gas cutting has an effect on the composition and tissue on both sides of the incision, and can cause deformation of the workpiece. The overall efficiency is low when the mass is cut[4]. Therefore, the traditional blanking method and high energy consumption blanking process already can not meet the requirement of the new type of industrial production[5-6]. It is of great significance to study the new blanking method. So, we need to carry out careful design and research on the blanking equipment. This paper discusses the rotary shear blanking machine, high-speed shear principle and the rotary cutting mechanism are used[7,8], flywheel every rotation can work twice for blanking, greatly improving the efficiency of blanking.

2 The working principle of rotary high speed blanking machine

The schematic diagram of the high speed rotary shear blanking machine is shown in Fig.1.

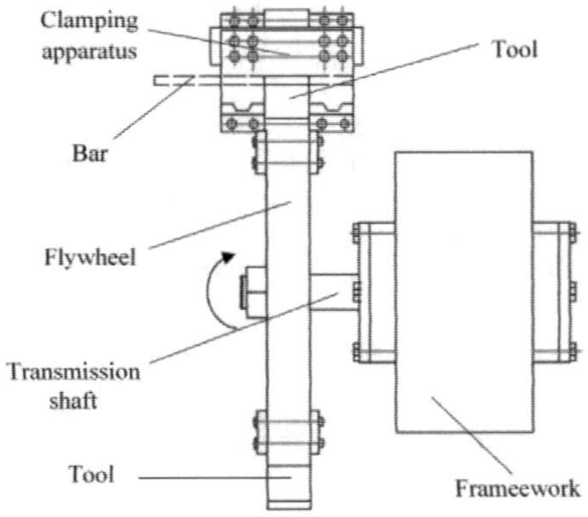

Figure 1 Rotary Shearing Principle Diagram

The rotary type high-speed shearing blanking machine includes the blanking mechanism, the clamping mechanism and the feeding mechanism, and the three parts together complete the cutting blanking of the metal bar. The blanking mechanism and the clamping mechanism directly affect the section quality of the metal bar, which is the core part of the rotary shear blanking machine. The feeding mechanism is the auxiliary mechanism. The main design of the blanking part is based on the diameter of the bar which needs to be cut.

In the cutting process, the bar is fixed with fixture, and the tool can be used for rotary shear. The bar is cut by relative movement of cutting tool and clamp. Gao[8] divided the shear process into elastic stage, plastic deformation stage, crack generation and fracture stages, we also divided into elastic stage, stage of plastic deformation, crack, and separation stage, as shown in Fig.2.

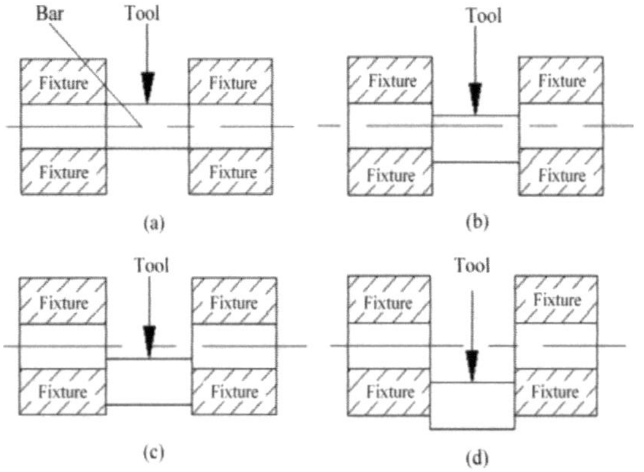

(a) Elastic Stage (b) Plastic Deformation Stage (c) Crack Formation (d) Separated Stage
Figure 2 High Speed Shearing Process Diagram

The basic working principle of the feeding machine is: the feeding mechanism adopts wheel transfer power, and the flywheel can complete the storage realization. The motor is driven firstly, the wheel transfer motion to complete the energy storage of the flywheel. The flywheel can achieve the function of the blanking by releasing energy. The clamping device is connected with the blanking platform by bolted connection. Rotary shear blanking machine need to cut the bar that with the same length and different diameter, that is why the mould clamping structure uses double V-type blocks. In order to replace the die easily, the traditional bolt connection is replaced by the hydraulic cylinder, and the stability and speed of clamping are guaranteed. The rod feeding mechanism is not very important in the whole shear system, but the accurate feeding is also an indispensable part of successful blanking.

3 Structure design of blanking system

3.1 The cutting tool structure

The structure design of the cutting tool is shown in Fig.3.

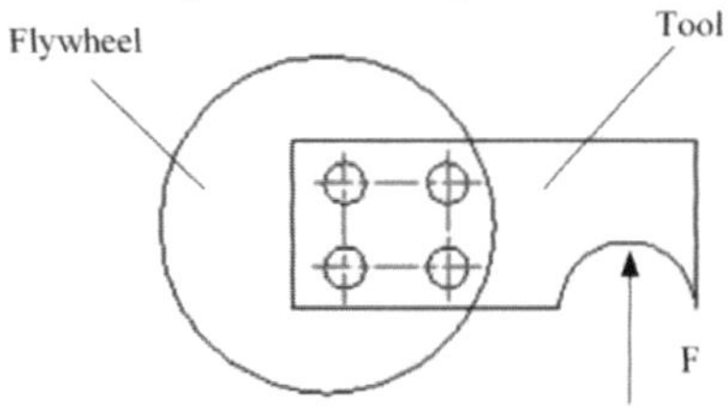

Figure 3 Tool Schematic Diagram

3.2 The fixture structure design

Because the rotary shear blanking machine need to blanking different types of metal bars, the mould uses V-shaped clamping piece to clamping, so that, it is convenient to replace the corresponding mold in the production of different types of bars. Adopts the form of radial convex platform fit into groove to realize axial fixation of the mould. A slot in the V-shaped block is open in order to connected to the radial of the hydraulic cylinder, bolted is used in axial connection to achieve the hydraulic cylinder fixing with V-shaped block. Same as the shape of a machine tool guideway is designed in the front of the V-shaped blocks, and cooperate with base hole corresponding to ensure V-shaped block only move along the radial direction, it can realize accurate clamping[9]. The mould can increase the contact area when the metal bar is blanking and reduce the deformation of the shear section on the metal bar. The convex platform in the center of the die is to prevent the axial movement of the mould. The shear mold structure is shown in Fig. 4.

Figure 4 Blanking Die

3.3 The shearing blanking system and feeding mechanism

In the initial state, the bars are arranged on a table with slope, and then two hydraulic cylinders rise at the same speed. The first bar will fall until the feeding block is intercepted and sent to the roller, during this period, the hydraulic cylinder will block the second rod falling. After the first rod is delivered, the hydraulic cylinder drops to the initial position, while the second rod is reduced to the position of the first bar. The fixed mode of hydraulic cylinder and hydraulic pump is the same as the clamping mechanism. But considering the speed of the hydraulic cylinder rising or falling too much, in the process of rising the bar avoiding pop-up deviates from the original set route, in setting speed regulating valve into the oil and out of the oil, guarantee hydraulic cylinder can quickly and stability lift and transport bar. The general assembly drawing is shown in Fig. 5.

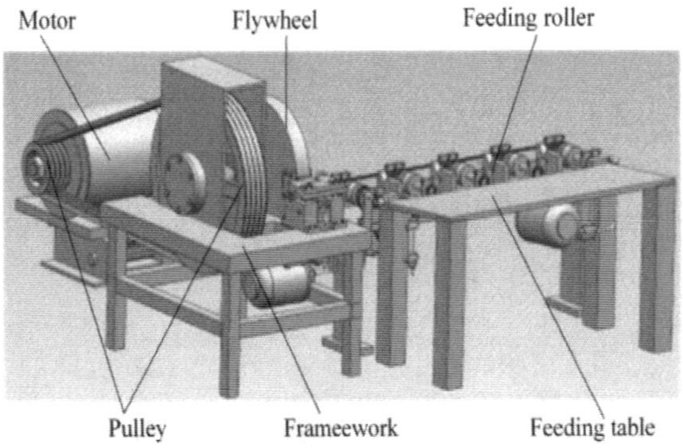

Figure 5 Shearing Blanking General Assembly

For the feeding mechanism, the whole feeding device how to accurately connect with blanking device and clamping device should to be considered. Based on this consideration, the height of the feeding device is adjusted by the friction between the bolted and similar keyway, ensure bar can accurate be sent to the clamping device. Rotary type high-speed shearing blanking system, including the feeding machine and feeding device.

4 Structure optimization of blanking tool

As the tool is the core part of the cutting blanking process, the strength must meet the design requirements. The finite element analysis of the cutter is shown in Fig.6.

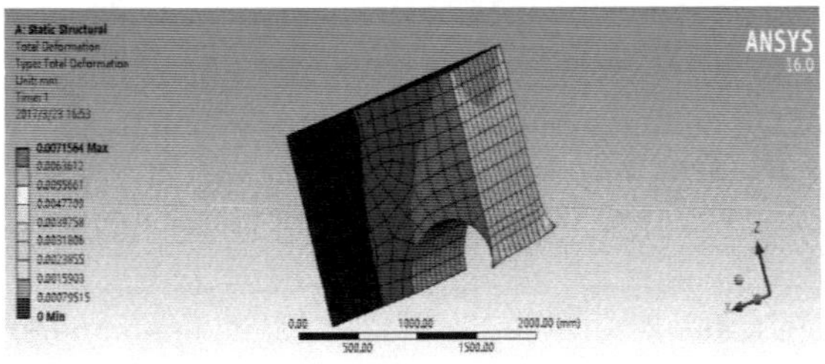

Figure 6 The Cutting Tool Stress Distribution Nephogram

From the Fig. 6, we can see that stress concentration is relatively serious in the lateral of tool opening, structure with large deformation and the tool has a larger damage. This is because the structural design of the tool is unreasonable and the shear force is too large in the blanking process. So should to increase the strength in the outside of the tooling opening, optimize the structure, re-establish the geometric model and mesh, the stress distribution cloud picture as shown in Fig.7. Stress concentration phenomenon of new structure greatly abate. The maximum stress is 0.3655MPa after optimizing, the stress value is reduced greatly, has been not enough to make the optimized cutting tool produces obvious deformation, and in range of the fourth intensity, therefore the optimization results can meet the strength requirement[9-10].

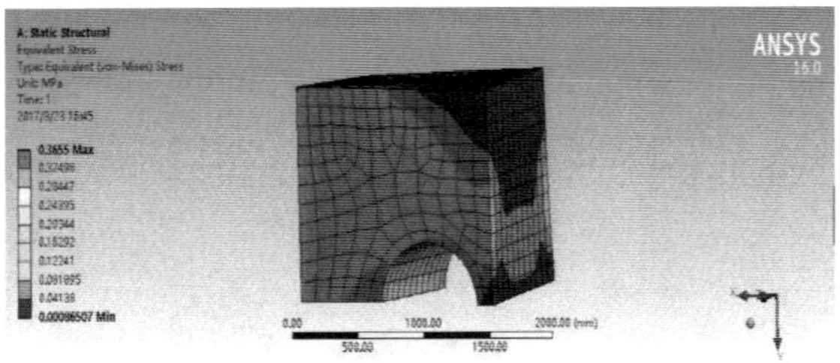

Figure 7 The Optimized Cutting Tool Stress Distribution Nephogram

5 Experimental study

5.1 Test equipment

Under 10 mm diameter in range of the small and medium-sized diameter bars, can be directly sheared by the rotary shear blanking machine[11-14], the picture of real products of the blanking machine as shown in Fig. 8.

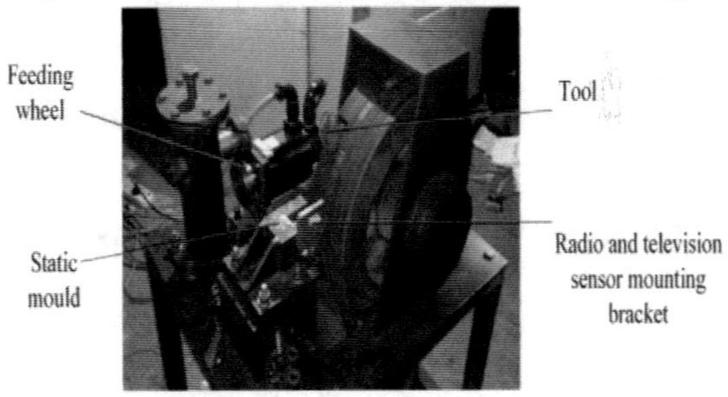

Figure 8 Pictures of the Rotary Type High-speed Shearing Blanking Machine

The feeding machine are consists of dynamic cutting edge, static mould, feeding wheel, clamping wheel, servo motor, main motor, photoelectric sensor and control system. The prefabricated circular V groove bar with certain geometric parameters is fixed on the holding wheel and the feeding wheel, the bar is feed by the servo motor in the signal drive of the control system, and the induction signal controlled by the photoelectric sensor signal control, each time the photoelectric sensor sensing dynamic shear blade, the servo motor feed the bar. In a set time, one of the dynamic shear blades just finishes shearing after the bar is feed.

5.2 Experimental results and analysis

Because the bar is in the single cantilever state in the process of shear, often caused some defects in the cross-section of bar, such as pits in one end and bumps in the other, as shown in no.2 test bar of Fig.9. Although the pit is not very deep, affect the blanking precision. In order to solve this problem, the state of bar is to be adjust, each time a bar is feed let the servo motor feed a more certain length, let the bar be constrained on both ends, at this time, doing the high speed shearing blanking, the bar getting from the blanking as shown in no. 3 sample in

Fig.9, and the section effect is improved. The best highest blanking speed is 150 piece/min.

Figure 9 Φ8mm Stainless Steel Bar Stock Cutting Section

The cross-sections of the traditional shear blanking methods and rotary type high-speed shearing blanking process are shown in Fig. 10. The quality of blank section can be clearly seen from Fig.10. It can be seen that and the quality of section Fig. 10 (c) is obviously better than that of Fig.10 (a) and Fig.10 (b). The cross-section obtained from the torsional shear blanking process is shown in Fig. 10(a). The tear phenomenon is obvious on the cross-section, there are large area raised and pits, and it is poor quality. In the shear blanking methods, the torsion shear efficiency is low, punching shearing blanking process rotary type high-speed shear blanking process are more efficient. The cross-section obtained from punching shearing blanking process is shown in Fig. 10(b). The profile of the cross-section is obviously deformed, with a horseshoe shape. The cross-section obtained from rotary type high-speed shearing blanking process is shown in Fig. 10(c). The billet is no deformation at the end, with smooth section and good quality.

(a) Torsion Shear Process (b) Punching Shearing Blanking Process (c) Rotary type high-speed shearing Blanking Process
Figure 10 Shearing Blanking Bar Section

Comparison with the punching shear process and rotary type high-speed shear blanking process, the test results of can be obtained as shown in Table 1.

Table 1 Test Data Comparison of Three Kinds of Shear Blanking Process

Blanking process	Bar's Materials	Bar's diameter	Height of bump	Height of pit	The percentage of defects	The geometry of the cross-section	Blanking speed
Torsion shear	304 stainless steel	Φ8mm	2.25mm	3.12mm	65%	Round shape	15 piece /min
Impact shear	304 stainless steel	Φ8mm	1.15mm	2.51mm	12.5%	Oval shape	60 piece /min
Rotary type high-speed shear	304 stainless steel	Φ8mm	0.35mm	0.41mm	0.5%	Round shape	150 piece /min

From Table 1, we can see that cross-sections of billets have the very high geometric accuracy, verticality and roughness. There are nearly no geometrical deformation and burrs, can be directly used in cold extrusion production such as gears, spline shaft and guide sleeve. However, the traditional punching shear process has the defects such as horseshoe shape, burr and so on. And the values of pits and bumps are relatively large. Especially the ratio that defects area occupies the whole section area reached 12.5%, it is too more. And the whole contour of cross-section got from the punching shear blanking of has passed from the initial circle to oval. It cannot be directly used in the next production procedure. Their cross-sections are shown in fig.10. Firstly defects should be treated, then it can be used for cold extrusion production such as gear, spline shaft and guide sleeve. At the same time, the rotary type high-speed shear blanking machine greatly reduces the loss of materials and also improves production efficiency.

In conclusion, the billet cross-section obtained from rotary type high-speed shearing blanking process has the very high geometric accuracy, verticality and smoothness. There are almost no defects such as geometric deformation and burrs. It can be directly used in cold extrusion production, such as gears, spline shaft and guide sleeve. However, the cross-sections got from the traditional shear blanking method have the defects such as horseshoe deformation or obvious bumps phenomenon and tear. This kind of billets with defects should go through the subsequent treatment processing. And then can be used for the cold extrusion production such as gear, spline shaft and so on. As a result, the new type of shear blanking process has greatly reduced the loss of materials and improved production efficiency.

6 Conclusion

By the related test and study showed that shear rate of 6 to 7 m/s in the high-speed rotary shear blanking process, be close to the critical brittle fracture speed of the metal, then quality of cross-section is best. Its accuracy is equivalent to the precision of the sawing machine, and retains some advantages such as die is simple and economical of the common shearing blanking.

(1) In this paper, a new type of rotary type high-speed shear precision blanking process is introduced, and the blanking machine structure and the feeding device have been designed and the experimental platform has been completed.

(2) Φ8mm diameter stainless steel bars were selected for experiment research, the blanking efficiency is improved and the good cross-section quality of billet is obtained. The largest convex and concave depth together with the defect proportion of cross-section are used to evaluate the quality of cross-section.

(3) The blanking technique is more suitable for medium and small diameter bar under the expected, the brittle material and plastic material for different bars should be to experiment research.

References

1. Li J W, Hu C L, Zhao Z, Li Z M .Influence of shearing speed on section quality for bar shearing[J]. Forging & Stamping Technology, 2016, 41(8):122-126.
2. Guo B T. Current situation and developing trend of global hacksawing machines[J]. Manufacturing Technology & Machine Tool, 2004,0(2):32-33.
3. Du S W. Bar high speed shearing machine[M]. Beijing: Metallurgical Industry Press, 2009.
4. Yu YP, Yang QH, Hu MJ. Numerical simulation on the blanking of metal tubes[J]. Mechanical Materials and Manufacturing Engineering, 2014, 455: 123-126.
5. Li Y H, Li Y T. Research and application of shearing speed of high-speed precision shearing for metal bar[J]. Hot Working Technology, 2010, 39(15): 8-11.
6. Hao B H. The analysis of blank shearing process and the methods for imporving breskan section quality[J]. Metal Forming Technology, 2000, 18(4):44-45.
7. Zhong W, Zhao S D ,Wang Z Y, Dong Y Z, Jing F. Discussion about method of improving the bar cutting surface quality through raising the hard brittle of material [J].China Metal forming Equipment & Manufacturing Technology,2016,51(02):77-83.
8. Gao L J. A metal bar material of high speed precision shearing experimental study[D]. Taiyuan: Taiyuan University of Science and Technology, 2008.
9. Cui Y B, Wang Z, Cui Y. The notice of using V type block[J]. Science & Technology Information, 2011,(9):27.
10. Tu Z F. ANSYS Finite element analysis of the engineering application instances tutoria[M]. Beijing: China Building Industry Press, 2010.
11. Tong Y, Zhao S D, Wang Z W. Development of a new type precision cropping machine with rotary striking[J]. China Metal forming Equipment & Manufacturing Technology, 2010, (06):14-17.
12. Yan H P. The designing about structure of cut equipment of high speed cut machine YAN Hui-ping[J]. Manofacturing Automation, 2002, 24(10): 26-32.
13. Tong Y, Zhao S D, Wang Z W. Experimental study on new precision cropping process for metal bars[J]. China Mechanical Engineering, 2010, 21(03): 359-363.
14. Zhao R F, Zhao S D, Jing F, Li J X, Guo T. Experimental study of new precision blanking technology for thick-wall tubes[J]. Forging & Stamping Technology, 2014, 39(8): 105-108.Here are some examples:

Acknowledgement

The authors wish to express gratitude to the Foundation of Shaanxi Province Educational Committee (Approval No. 16JK549) and Natural Science Foundation of Jiangsu Province (Approval No. BK20140407) which have supported this work.

Neural Network Optimized Model Predictive Multi-Objective Adaptive Cruise Control

Siyi Zhang[1], Junzhi Zhang[1,2]

[1]*State Key Laboratory of Automotive Safety and Energy, Tsinghua University, Beijing, People's Republic of China*

[2]*Innovation Center of Electric Vehicles, Beijing, People's Republic of China*

Abstract. A model predictive multi-objective adaptive cruise control (MPC MO-ACC) system, designed to consider both the tracking performance and the fuel consumption, is optimized by a neural network in this paper, reducing the computational complexity without sacrificing the control performance. The optimized MO-ACC control system is built by training a neural network with the control results of the MPC MO-ACC system. Simulation tests are conducted in Matlab/Simulink in conjunction with the high-fidelity CarMaker software. Influences of four driving conditions (the learning track, NEDC, JP05, FTP75) and two kinds of sensor models (ideal radar sensor and 77GHz physical radar sensor) are analysed. Simulation results have shown that the neural network optimized model predictive MO-ACC has the same control capability and strong robustness as the original MPC MO-ACC. Meanwhile, the optimized control system has much lower computational complexity, which shows potentials for the application in real-time vehicle control and industry.

1 Introduction

With the progress of intelligent vehicle control, the normal Adaptive Cruise Control (ACC) system, which has the basic functions such as cruise control and active collision avoidance[1], has developed into the Multi-objective Adaptive Cruise Control (MO-ACC) system, which considers not only the tracking performance but also the fuel consumption, and driver's comfort as well[2].

Reference [2] developed an MO-ACC system based on Model Predictive Control (MPC). It is proved that MPC can make multi-objective optimization simultaneously, and has the features of good control performance and high robustness. However, the high computational complexity of MPC has restrained its application in real-time vehicle control.

There are two main solutions to solve this problem: 1. improving the hardware computational ability, but will increase costs as well; 2. building an off-line look-up table, but workload of building such a table is very high.

Meanwhile, neural network has been implemented in ACC systems. Reference [3] has built a neural network ACC system on the base of human drivers' control results provided by NGSIM, which can ensure a safe headway for vehicles under most circumstances. However, fuel consumption was not considered in such kind of neural network ACC systems.

This paper has shown an optimization of the MPC MO-ACC system by using the neural network. The optimized MO-ACC system is built by training a neural network with the control results of the MPC MO-ACC system, reducing the computational complexity without sacrificing the control performance. Compared to the previous methods, this optimization is more cost-efficient and labour-efficient.

This paper is organized as follows. In Section 2, the training process of the network is introduced. In Section 3, results of the simulation conducted in Matlab/Simulink in conjunction with the high-fidelity CarMaker are analysed. The influences of different driving conditions and the noise of physical radar sensors are also considered in the analysis. In Section 4, the advantages of this neural network optimized model predictive multi-objective adaptive cruise control system are concluded.

2 Training of the neural network

The neural network is trained with the control results of the MPC MO-ACC, when ideal sensors are used. The vehicle speed, acceleration and headway of the preceding and following vehicles, and the acceleration command value of the previous step are selected as inputs of the control system, then the acceleration command value of this step is calculated as the output.

Series of experiments are done to compare the influences of different network structures (fitnet, cascadeforwardnet and feedforwardnet), network nodes (10, 15 and 20 nodes), activation functions (tansig, puelin) and training functions (taingd, traingdm, traingdx, trainrp, traincgf, traincgp, traincgb, trainscg, trainbfg, trainoss, trainlm, trainbr) on the control performance. Each trial

 https://doi.org/10.1051/matecconf/201816601009

was repeated 5-10 times to ensure the reliability of the results.

It is shown that with the network structure of multi-layer feedforward neural network, one hidden layer, ten nodes, activation function 'tansig' for hidden layer, activation function 'puelin' for output layer, and training function 'trainlm' or 'trainbr', neural network control systems with good performance can be trained. The settings of the neural network used in this paper are as shown in Table 1.

Table 1. Settings of the neural network control system.

Network Element	Setting
Network structure	Multi-layer feedforward neural network
Number of hidden layers	1
Number of nodes	10
Activation function of hidden layer	tansig
Activation function of output layer	puelin
Training function	trainlm

3 Co-simulations and analysis

Co-simulations are conducted in MATLAB/Simulink and IPG CarMaker to compare the control performance of the original MPC and the neural network control system. The vehicle data of Honda Fit CVT provided by CarMaker 6.0.2 is chosen to build the vehicle dynamic model. An ideal radar sensor and a 77GHz physical radar sensor in CarMaker are used respectively to detect the relative distance, relative vehicle speed, and relative longitudinal acceleration. Together with the motion signals of the following vehicle, these signals are then calculated to get the inputs of the MO-ACC systems.

The MPC and neural network MO-ACC systems are evaluated by the tracking performance and the fuel consumption. To reflect tracking capability, the tracking error index (TEI) similar to that was defined in Reference [4] is used. It considers both speed error and distance error, as shown in Equation (1).

$$TEI = \frac{1}{T}\int_0^T (|\Delta d\,(t)| + 0.1|\Delta v\,(t)|)dt \qquad (1)$$

Where T is the simulation time, Δd is the difference between the desired distance and the actual distance in every step. Δv is the difference between the desired and actual velocity in every step. The units of Δd and Δv are m and m/s respectively.

The fuel consumption is reflected by the average fuel consumption per hundred kilometres (FCM). It is calculated as Equation (2).

$$FCM = \frac{1}{S}\int_0^T Q_{eng}\,dt \qquad (2)$$

Where Q_{eng} is the fuel consumption rate of the engine, whose unit is L/s. S is the total driving distance of the following vehicle, whose unit is 100km.

3.1 Simulation results using ideal radar sensors

Firstly, ideal radar sensors are used in the simulation, which means the relative velocity, relative distance and relative acceleration are very accurate, and so the control will not be influenced by the noise of sensors.

An MPC MO-ACC system is then tuned under the track shown in Figure 1. The target velocity and target headway, shown in black dashed lines, are tightly followed by the actual velocity and actual headway, which are shown in blue lines. The relative velocity between the preceding vehicle and the following vehicle is controlled within $\pm$1m/s, and the relative distance between the target and actual headway is controlled within $\pm$0.5m, which can show the good performance of the MPC control system.

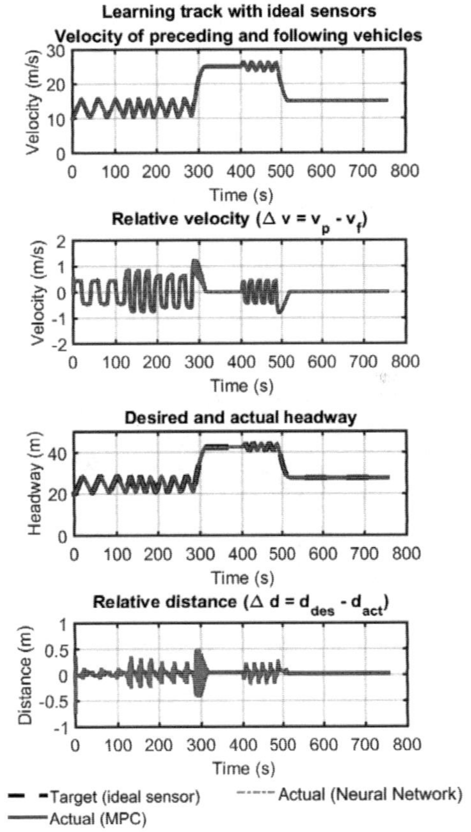

Figure 1. Simulation results under learning track with ideal sensors

After that, a neural network MO-ACC system is trained by the control results of the MPC system under the learning track with ideal sensors. The control results of this neural network MO-ACC system is shown in red lines in Figure 1, which highly overlaps that of the MPC system. This can be explained by the relative desired acceleration shown in Figure 2, which means the difference between the desired accelerations calculated by the MPC and the neural network control. This value is constantly within $\pm 3 \times 10^{-4}$m/s^2, which has almost no influence in the following vehicle dynamic control.

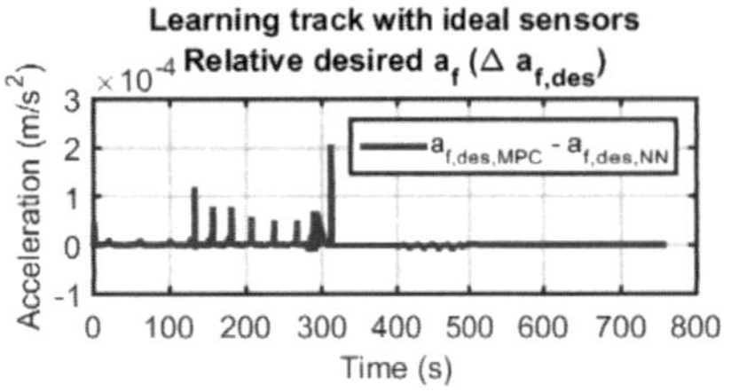

Figure 2. Relative desired acceleration under learning track with ideal sensors

Moreover, the original MPC and the neural network optimized control systems are tested under different typical worldwide driving conditions, including NEDC for Europe, JP05 for Japan, and FTP75 for the USA.

The control results shown in Figure 3 illustrate the good performance and strong robustness of both the MPC and neural network control. The relative velocity and relative distance are always controlled within ±3m/s and ±1m correspondingly.

Furthermore, the tracking performance and fuel consumption of these two control systems under different driving conditions are integrated in Figure 8. The TEI and FCM values of the neural network control are totally the same as that of the MPC when using ideal sensors. This can be explained by the slight relative desired acceleration between the MPC and neural network control system, as shown in Figure 4, which is constantly within $\pm 1 \times 10^{-3}$m/s^2. When considering the complicated dynamic control progress in the lower controller and the nonlinear characteristics of the vehicle body and tyres, it is understandable that the influence caused by such a small difference in the acceleration command is too hard to observe.

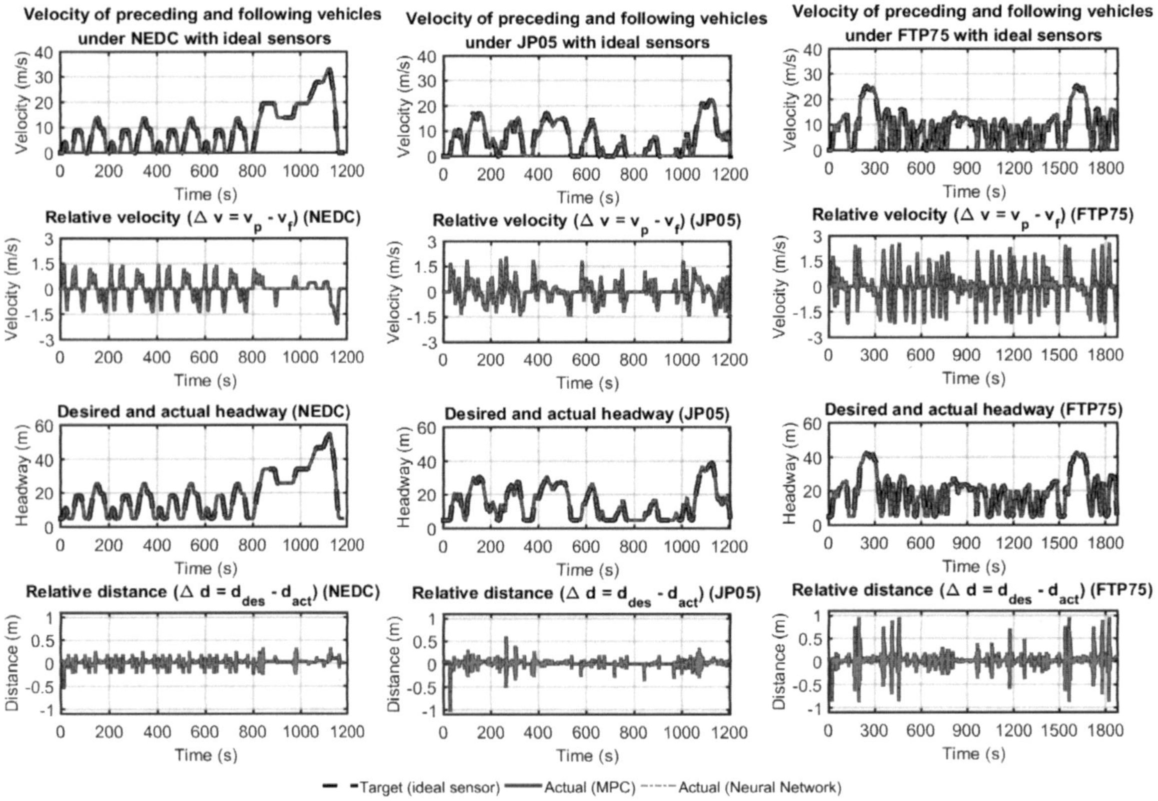

Figure 3. Simulation results under NEDC, JP05 and FTP75 with ideal sensors

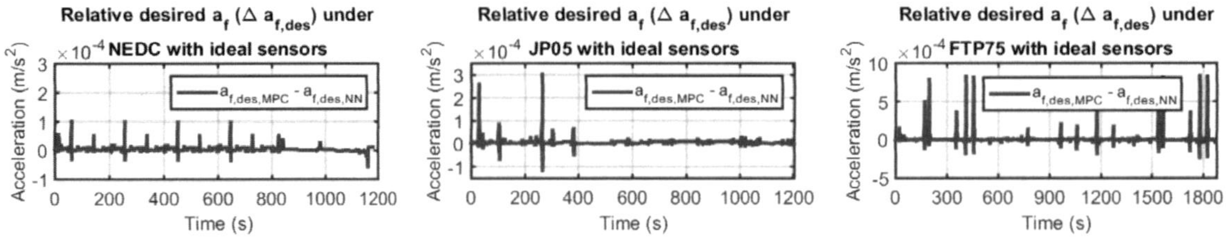

Figure 4. Relative desired acceleration under NEDC, JP05 and FTP75 with ideal sensors

According to the tests so far, it can be concluded that, when not considering the noise of sensors, the neural network MO-ACC system has not only the same good control ability but also the high robustness as the MPC system.

3.2 Simulation results using physical radar sensors

In this subsection, 77GHz physical radar sensors are used in the co-simulation, which means the noise of sensors will be considered in the following tests. Due to the randomness of the noise, all the tests presented here are repeated for 5 times to ensure the reliability of the results. Meanwhile, TEI is still calculated by the accurate relative velocity and distance, in order to eliminate the influence of the sensor noise on these evaluation indexes.

The noise bandwidth of radar is set to be 25000Hz and the noise figure is set to be 4.8dB, which are recommended default values in CarMaker. The influence of these noises can be seen apparently in the enlarged parts in Figure 5 and Figure 6. The relative velocity and actual distance between the preceding and following vehicles are measured by the physical sensors with noise. As a result, the velocity of the preceding vehicle and the actual headway will flutter around the accurate values given by the ideal sensors.

Even in this case, both the MPC and neural network control systems are still performing well. As shown in Figure 7, the relative velocity and relative distance are still controlled within ± 1.5m/s and ± 1m, demonstrating the good control ability and high robustness of both control systems.

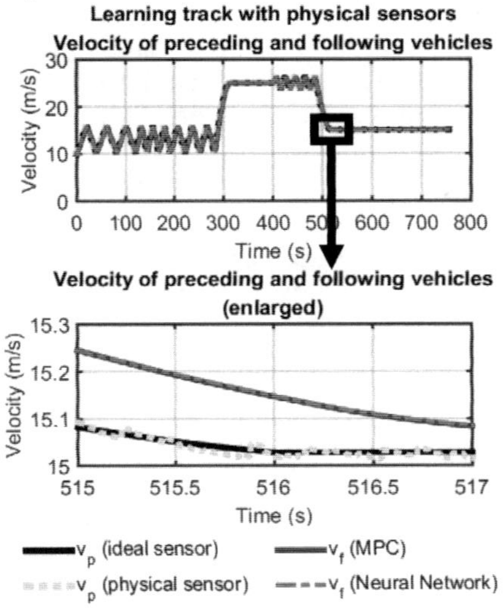

Figure 5. Velocity of preceding and following vehicles under learning track with physical sensors

However, the relative distance controlled by neural network is slightly higher than that by MPC. This can also be verified by TEI in Figure 8. When considering the noise of sensors, TEI of the neural network control system is slightly worse than that of MPC, where relative errors are 2.07% under the learning track, 2.80% under NEDC, 7.16% under JC05 and 6.86% under FTP75. However, since absolute errors are all below 0.01, the tracking performance of the neural network MO-ACC control can still be viewed as good as that of the original MPC system.

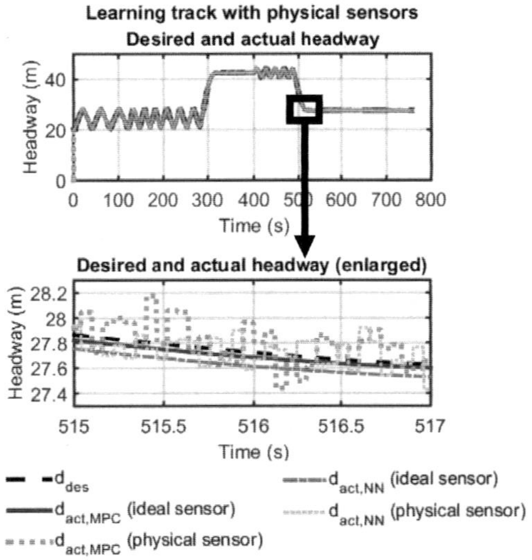

Figure 6. Desired headway and actual headway under learning track with physical sensors

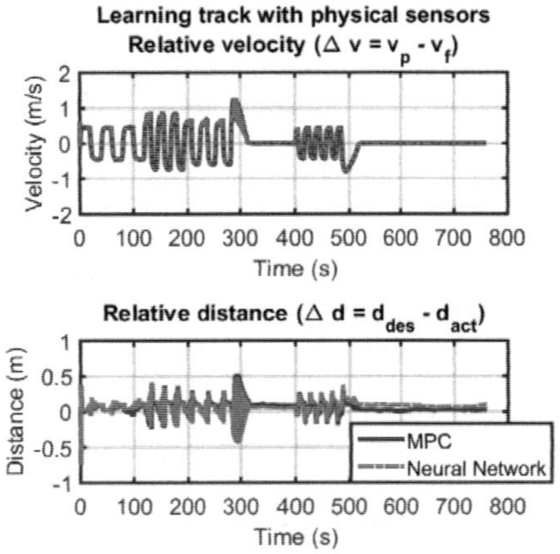

Figure 7. Relative velocity and relative distance under learning track with physical sensors

Moreover, the fuel consumption of vehicles controlled by different control systems under different driving conditions and with different kinds of sensors are shown in Figure 8. It can be seen that the influence of sensors' noise is very small, and the performance of the original MPC and the optimized neural network control system are very similar. The error between the FCM of different sensors (ideal and physical radar) and the error between the FCM of MPC and neural network control systems are all below 0.01L/100km. When tested under NEDC with physical sensors, a vehicle controlled by the neural network consumes even less fuel than a vehicle controlled by an MPC system. Thus, it can be proved that the neural network MO-ACC control system has the same control capability and strong robustness as the original

MPC system, in aspects of both the tracking performance and fuel consumption.

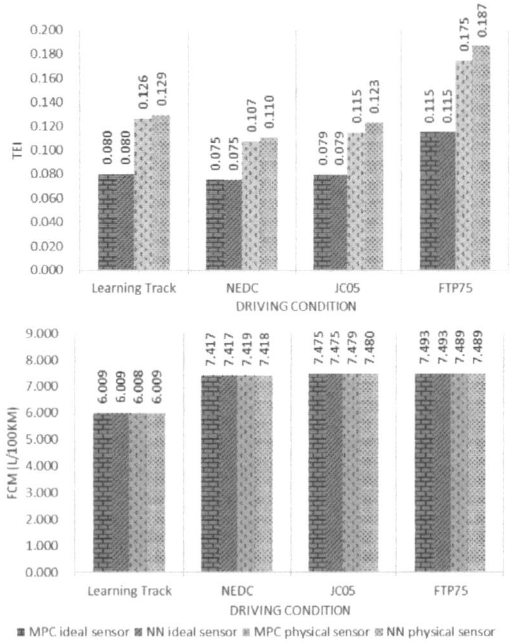

Figure 8. TEI and FCM comparison of MPC and Neural Network control systems with ideal and physical sensors

3.3 Comparison of the computational complexity

After comparing control capability and robustness, the computational complexity of these two control systems are analysed in this subsection, which is measured by the computing time on the same hardware. Simple lower controller and vehicle dynamic models are used here, in order to make sure that the computing time is only influenced by complexity of MO-ACC control systems.

As shown in Table 2, the computing time of the neural network control system is more than 300 times shorter than that of the MPC system. This means by using the neural network control, the computational complexity can be reduced by more than 300 times, which is a great feature in real-time vehicle control and has a great potential for the wide application in industry.

Table 2. Comparison of the computing time of the MPC system and the neural network control system

	Time using MPC (s)	Time using neural network control (s)	Multiple
Learning track	1200	4	300
NEDC	2000	6	333
JP05	2040	6	340
FTP75	3180	10	318

4 Conclusion

A new method to optimize model predictive MO-ACC systems is given in this paper. After being trained with the control results of the MPC MO-ACC system, the neural network MO-ACC system shows almost the same good control ability and strong robustness as the original MPC system. Meanwhile, the computational complexity of the MPC system can be reduced by more than 300 times in this optimization, which makes its application in real-time control possible.

This MPC-trained neural network MO-ACC system considers not only the tracking performance, but also the fuel consumption, which is an advantage over the normal neural network control systems trained by human drivers' control behaviour. Besides, the training process requires no big data and data cleaning. Furthermore, control systems which can consider more factors can also be designed in this method.

Moreover, compared with other methods to reduce the computational complexity of MPC, such as updating the hardware or building a large off-line look-up table, the optimization method given in this paper is more cost-efficient and labour-efficient. As a result, the neural network optimized model predictive MO-ACC system features not only the good control ability and strong robustness, but also the low computational complexity and low cost, which is believed to have a great potential for the application in real-time vehicle control and industry.

Funding

This work was supported by the National Key Research and Development Program of China [project no. 2016YFB0101402]; the National Natural Science Foundation of China [project no. 51475253].

References

1. J. Zhang, Z. Liu, Z. Qi, et al., Development of Automotive Adaptive Cruise Control System, Vehicle & Power Technology **2**, 44-47 (2003)
2. S. Li, J. Wang, K. Li, et al., Processing of MPC practical problems and its application to vehicular adaptive cruise control systems, J Tsinghua Univ (Sci and Tech) **50**, 645-648 (2010)
3. W. Shen, The Research on Adaptive Cruise Control System Based on Neural Network and PID, Jiangsu University (2015)
4. S. Li, K. Li, R. Rajamani, et al., Model Predictive Multi-Objective Vehicular Adaptive Cruise Control, IEEE Transactions on Control Systems Technology **19**, 556-566 (2011)

Estimation of Vehicle Side-Slip Angle Using an Artificial Neural Network

Daniel Chindamo[1] and Marco Gadola[1]

[1]*Department of Industrial and Mechanical Engineering, Automotive Group, University of Brescia, I-25123 Brescia, Italy.*

Abstract. In this work, a reliable and effective method to predict the vehicle side-slip angle is given by means of an artificial neural network. It is well known that artificial neural networks are a very powerful modelling tool. They are largely used in many engineering fields to model complex and strongly non-linear systems. For this application, the network has to be as simple as possible in order to work in real-time within built-in applications such as active safety systems. The network has been trained with the data coming from a custom manoeuvre designed in order to keep the method simple and light from the computational point of view. Therefore, a 5-10-1 (input-hidden-output layer) network layout has been used. These aspects allow the network to give a proper estimation despite its simplicity. The proposed methodology has been tested by means of the CarSim® simulation package, which is considered one of the reference tools in the field of vehicle dynamics simulation. To prove the effectiveness of the method, tests have been carried out under different adherence conditions.

1 Introduction

It is well known that vehicle side-slip angle is the key to understand vehicle lateral dynamics. It can be used to investigate the cornering behaviour in terms of understeer/oversteer and stability, which are related to active safety, driveability and performance, either for everyday road cars and high-performance vehicles [1, 2]. Moreover, vehicle side-slip angle is important to feed and tune active safety systems such as electronic stability program (ESP, see Table 1 for a comprehensive nomenclature), dynamic yaw control (DYC) and antilock braking system (ABS) as well as other active features like electronically controlled differentials.

Devices required to obtain a direct measure of the vehicle side-slip angle are expensive and generally difficult to fit on a car. Basically, such devices are optical transducers requiring extremely precise alignment in order to give a reliable measurement. They are also very susceptible to road surface type and weather conditions.

With regard to motorsport, racing rules often make on-board installation forbidden.

Therefore, the side-slip angle has to be estimated on the basis of conventional measurements carried out on the vehicle, i.e. lateral/longitudinal acceleration, yaw rate, steer angle and angular speed for each wheel, allowing to estimate the vehicle speed. Several side-slip angle estimation strategies have been proposed in literature. They are mainly based on Extended Kalman Filters (EKF) [3-5]. Since EKF typically requires a reference vehicle model, it is able to provide a good estimation only if model parameters are accurately known. Under this point of view, tyre characteristics are particularly critical depending on various factors like road surface conditions, temperature, tyre wear, etc.

Unfortunately, many of the parameters and data required to build the reference vehicle model are often unknown or difficult to measure, especially those referred to a key component such as the tyres. In fact, obtaining tyre data (i.e. Pacejka coefficients) from tyre manufacturers is often a real challenge.

Hence, in this paper the need of a deep vehicle knowledge has been avoided by using an artificial neural network (ANN) instead. In fact, using ANN the vehicle can be considered as a black box system and only a conventional set of sensors is needed to train/feed the network.

Artificial neural networks are considered effective tools for system modelling. Neural networks also offer decisive advantages such as adaptive learning, fault tolerance and generalization. Moreover, in recent years, the development of high speed computers encouraged the application of artificial neural networks which have been progressed very quickly.

Thus, among the others, ANNs have been largely applied also in technical disciplines like automotive engineering.

This method has been used, among others, in [6] to build a hybrid vehicle power management strategy, in [7] to model Diesel engine gas emissions, in [8] to estimate the position of an autonomous vehicle, in [9] for a longitudinal wheel slip controller, in [10] to monitor a Racecar driver vital parameters.

Therefore, literature shows that artificial neural networks are suitable to model complex systems using their ability to identify relationships from input-output data [11].

Some attempts of applying a neural network technique to vehicle side-slip angle estimation can also be found in literature. In [12, 13], the side-slip angle is estimated as a function of lateral acceleration and yaw rate. The results seem to be promising, but the effects of vehicle speed variations hence tyre–road friction are not addressed. The neural network suggested in [14] is instead based on a non-conventional set of sensors (measurements provided by four two-axis accelerometers placed on each corner of the car body are required).

Moreover, promising results have been obtained also in [15] and [16]. However, a previous estimation of the side-slip angle as input to the network (feedback) along with a large number of inputs (i.e. 9 inputs leading to many more calculations), are needed to let the ANN work properly. Hence, more calculation and CPU usage are required to deal with the network complexity.

Therefore, the aim of the present work is to further investigate the application of an ANN in order to develop a lighter solution able to give a reliable and effective real-time estimation of the vehicle side-slip angle.

ANNs give better results as much as the sets of data used to train the network contain the whole possible scenario the system may have to deal with [17], thus a specific training procedure for this application has been also proposed to keep the network layout as simple as possible hence light from the computational point of view. This allows to use a network with a very low number of neurons (5-10-1 input-hidden-output layer), reducing the number of calculations needed. This is crucial for real-time operation.

Indeed, a validation of the proposed methodology has been provided. It is carried out by means of the CarSim® package which is considered one of the reference tools in professional vehicle dynamics simulation. In order to simulate data coming from real-world sensors, CarSim® outputs have been added with uncertain random noise variance as successfully used in [18].

As reported in section 4, the proposed methodology leads to good results while keeping the network simple. The computational time is very low indeed therefore suitable for real-time applications.

It is worth to stress that this method takes into account the vehicle speed variation and is conceived to be fed using a simple and cheap set of sensors usually already onboard of the most recent passenger cars.

2 Methods

The approach used in this work is the same adopted by Parlak et al. in [19]. It uses the Levenberg-Marquardt back-propagation algorithm along with the mean squared error (MSE) performance index [20]. This BP optimization algorithm updates weights and biases of the network as presented in [21, 22]. See Figure 1 for the high-level schematics.

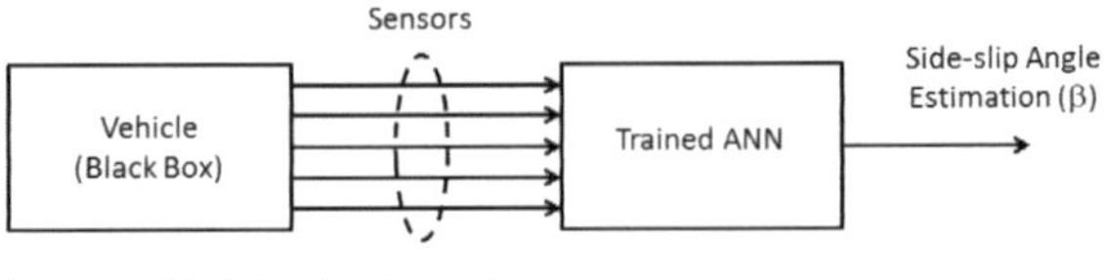

Figure 1. High-level schematics.

Then, the ANN performance has been determined by the MSE which is:

$$MSE = \frac{1}{N}\sum_{i=1}^{N}\left(y_i - y_k\right)^2, \qquad (1)$$

where y_i and y_k are respectively the predicted value and the target value of the i-th pattern and N is the number of pattern.

The dataset required to train and test the ANN has to be composed by input data patterns together with corresponding targets. Moreover, it has to be divided in two different sets, in this case 90% of the dataset has been used to train the network while the remaining 10% has been used to test the capability of the network. This is a common procedure as reported in [23].

Required mathematical background along with testing and training procedures can be found in [24].

For this application inputs (which determine the ANN's input layer neuron number) are: vehicle speed (vx), steering angle (δ), lateral acceleration (ay), longitudinal acceleration (ax) and yaw rate (ψ). The output is the vehicle side-slip angle (β), hence the ANN has only one neuron in its output layer.

As stated above, this estimator is supposed to be reliable and light from the computational point of view, thus it can be used in real-time built-in applications such as vehicle active safety systems (i.e. ESP, DYC, ABS, etc.). Therefore, the neuron number of the hidden layer, which leads to satisfy both accuracy and simplicity purposes, is 10. It has been chosen using an iterative "trial and error" approach.

The resulting 5-10-1 (input-hidden-output layer) network layout has been built using the "Neural Networks Toolbox" of MATLAB®. Note that, for hidden and output layer the log-sigmoid transfer function has been used.

Finally, the ANN's structure is reported in Figure 2.

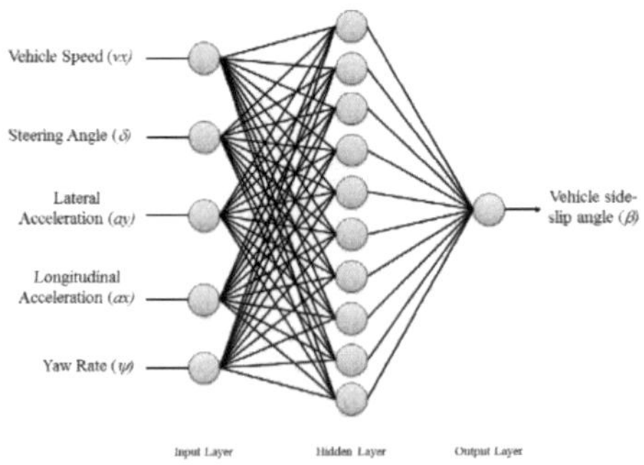

Figure 2. Layout of the adopted ANN.

 https://doi.org/10.1051/matecconf/201816602001

2.1 Training data set and procedure

The most challenging part of this work was the identification of an appropriate manoeuvre to be used for the ANN training.

As stated above, the training data set must represent as faithfully as possible the system to be simulated and its dynamics in all conditions. When this criterion is met, the neuron number of the hidden layer can be reduced (within certain limits), giving a simple and light ANN.

It is worth stressing that, for this application both the tyre linear and non-linear behaviours must be taken into account, along with the vehicle behaviour as a function of its speed. Therefore, a custom manoeuvre that explores the whole tyre range of usage has been designed.

In order to take into account the whole range of tyre slip angles, a sinusoidal steer manoeuvre has been adopted.

Hence, the proposed training manoeuvre is a 45° @ 0.25 Hz sinusoidal steer carried out while the vehicle accelerates following a 0.25 km/h/s ramp from 5 km/h to 100 km/h, i.e. until tyre saturation is reached.

In order to deal with all the possible meteorological conditions, such manoeuvre has been carried out with different tyre-surface friction coefficients (μ). So that, training dataset is composed by the same manoeuvre repeated for each μ (from 0.4 simulating sleet/snow conditions to 1 simulating dry surface with step increment of 0.1).

Indeed, the data obtained this way allows to meet the above-mentioned criteria. Moreover, in a real-world context, this manoeuvre can be easily performed by a well-trained test driver on any proving ground.

Figure 3 represents the vehicle speed and steering wheel data histories (see Figure 3a and 3b) of the training manoeuvre. Also the resulting vehicle side-slip angle for μ=1 has been reported for reference (see Figure 3c).

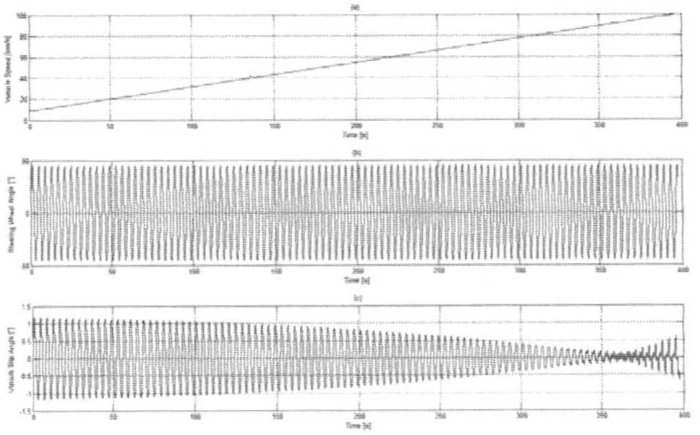

Figure 3. Adopted training manoeuvre. (a) vehicle speed, (b) steering wheel, (c) vehicle side-slip angle for μ=1.

The Levenberg–Marquardt algorithm, fed with the training data set, has been able to train the network after 899 iterations, giving a final performance index (MSE) of 9.24e-6 with a gradient of 6.82e-5.

Therefore, the network can be considered ready to be tested with other generic data sets, as reported in the next section.

3 Results and discussion

As stated before, the proposed ANN has been trained and tested with a professional vehicle dynamics simulation software called CarSim® which represents a good reference in the vehicle simulation field. Many simulations were performed and compared to the results given by the network.

A GT class vehicle has been used for this purpose. Thus, a series of double lane change (DLC) manoeuvres have been simulated to check if the trained ANN was able to reproduce the vehicle lateral transient behaviour at different speeds. Also, a series of step steer inputs (SS) at different speeds have been simulated to further investigate the ANN capability and to make sure that vehicle steady state behaviour can be reproduced as well. Both DLCs and SSs have been carried out at 30 km/h, 60 km/h and 90 km/h. All the tests have been repeated using μ=1 simulating dry conditions, μ=0.7 simulating heavy rain conditions and μ=0.4 simulating sleet/snow conditions. It is worth stressing that, apart for the training procedure, the vehicle side-slip angle is not used as input value but just as reference. Hence, there is not feedback lines and the vehicle side-slip angle does not appear among the inputs in any form.

Test results for μ=1 are reported below in figures from 4 to 9 and they show that the trained ANN gives a very good estimation especially regarding manoeuvres performed at speed below 90 km/h where the mean error has been found to be less than 2%, especially in figures 4 and 7 where the difference between the estimated and actual value of the side-slip angle is hardly visible. However, figures 6 and 9 show a good estimation for manoeuvres performed at 90 km/h as well.

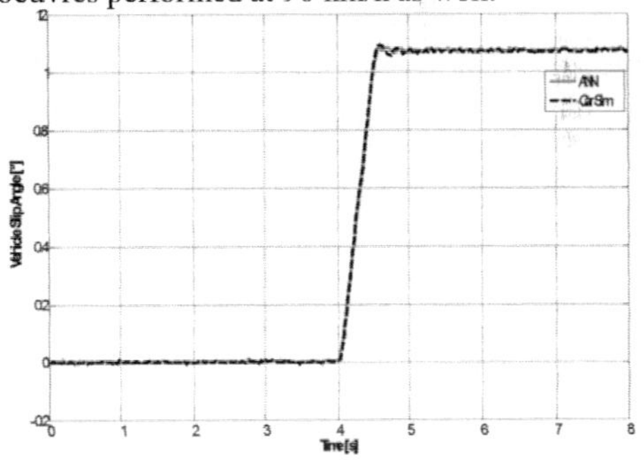

Figure 4. 45° step steer @ 30km/h, μ=1.

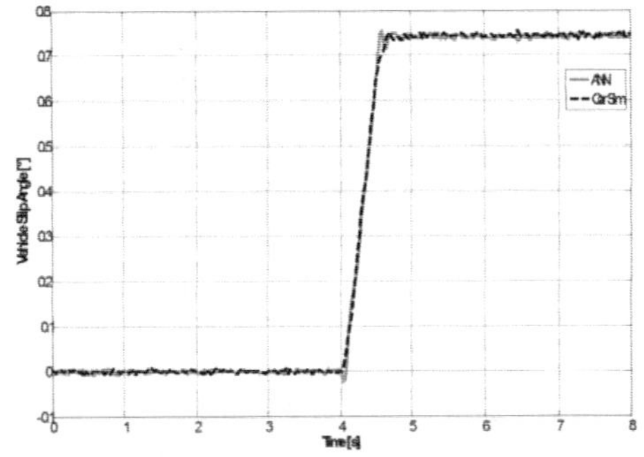

Figure 5. 45° step steer @ 60km/h, μ=1.

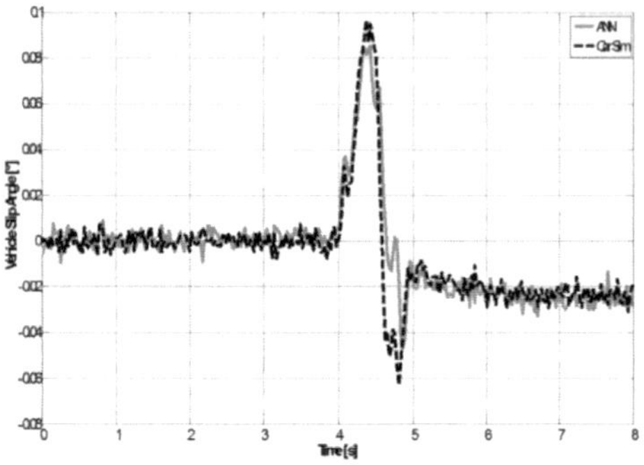

Figure 6. 45° step steer @ 90km/h, μ=1.

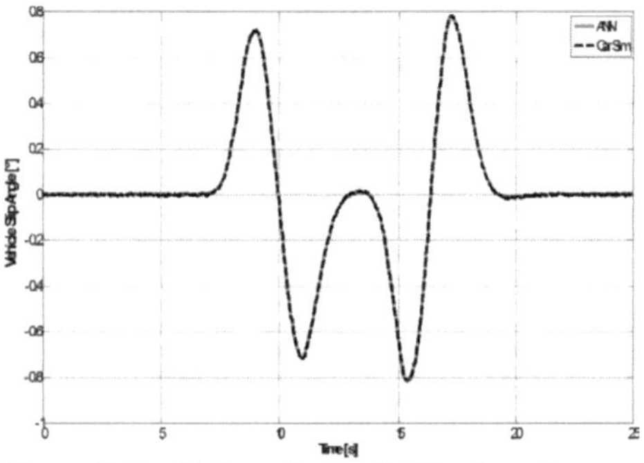

Figure 7. Double lane change @ 30km/h, μ=1.

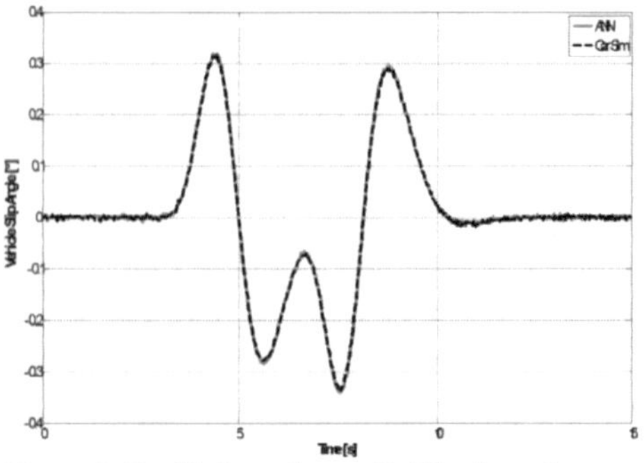

Figure 8. Double lane change @ 60km/h, μ=1.

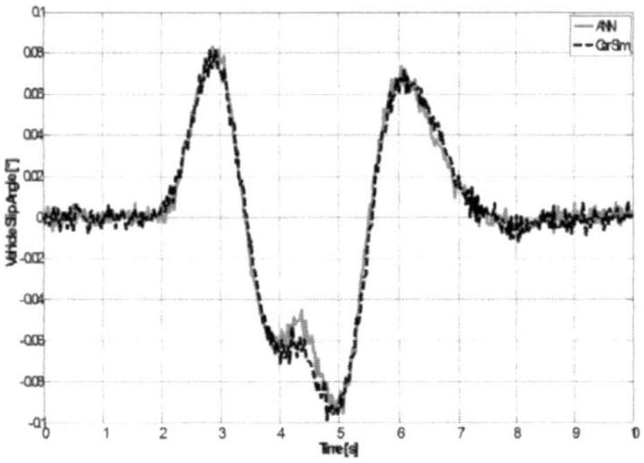

Figure 9. Double lane change @ 90km/h, μ=1.

Test results for μ=0.7 are reported below in figures from 10 to 15. Test results show that mean and maximum errors are always under 5%. Once again, best results are in the lower speed range hence with low tyre slip angles.

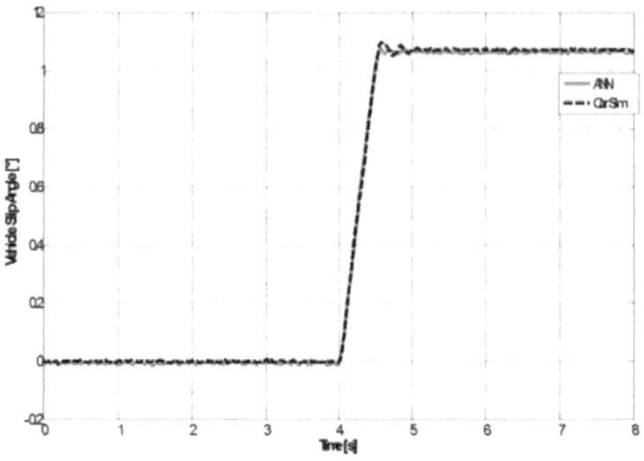

Figure 10. 45° step steer @ 30km/h, μ=0.7.

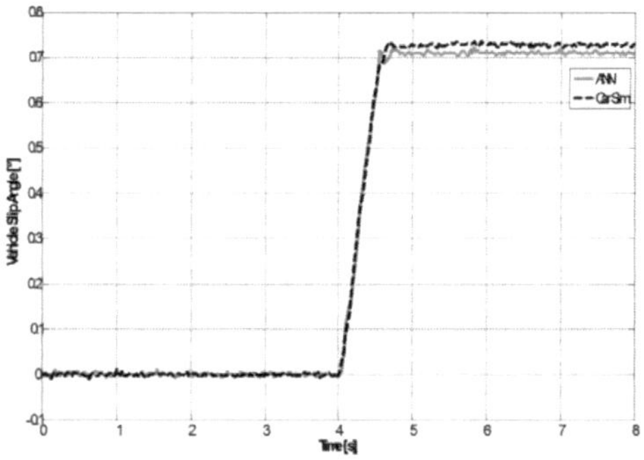

Figure 11. 45° step steer @ 60km/h, μ=0.7.

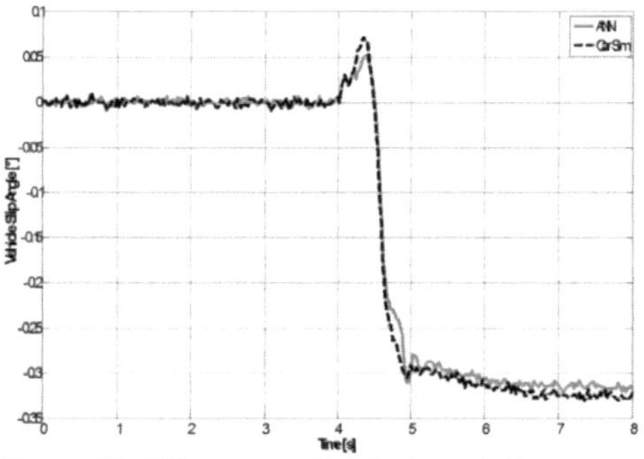

Figure 12. 45° step steer @ 90km/h, μ=0.7.

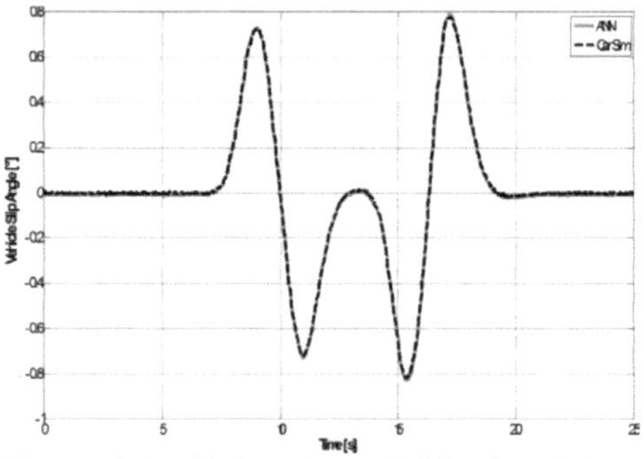

Figure 13. Double lane change @ 30km/h, μ=0.7.

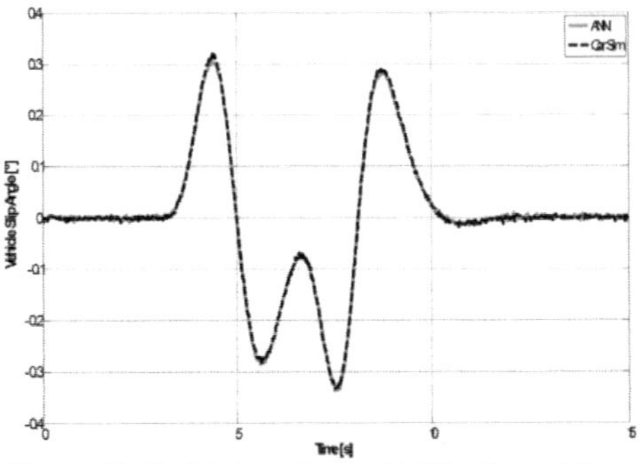

Figure 14. Double lane change @ 60km/h, μ=0.7.

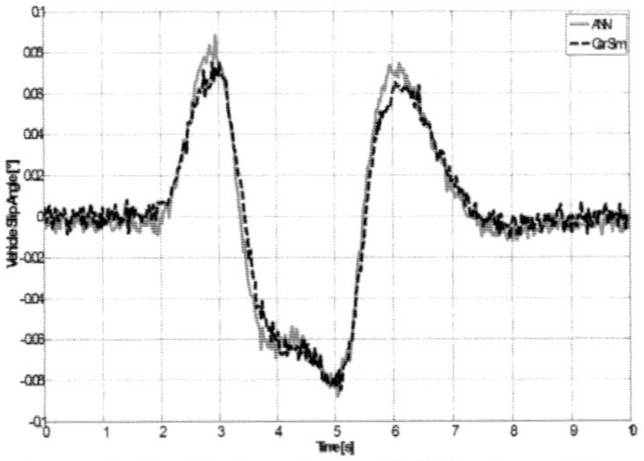

Figure 15. Double lane change @ 90 km/h, μ=0.7.

Test results for μ =0.4 are reported below in figures from 16 to 21. Note that, due to the low adherence conditions, tests have been carried out at 30 km/h, 40 km/h and 50 km/h. Also in this case, both mean and maximum error between estimated and actual value of the vehicle slip angle are well below 5% in the lower speed range (see figures 16, 17, 19, 20).

However, tests performed at 50 km/h show bigger errors (se figures 18 and 21). It is worth highlighting that a double lane change and step steer performed at 50 km/h wit μ=0.4 are very challenging manoeuvres.

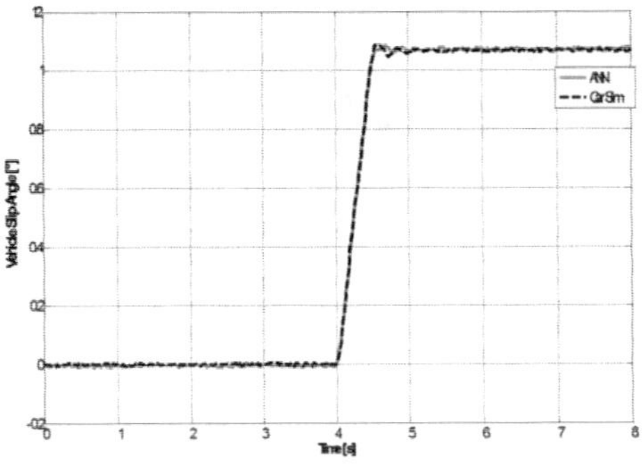

Figure 16. 45° step steer @ 30 km/h, μ=0.4.

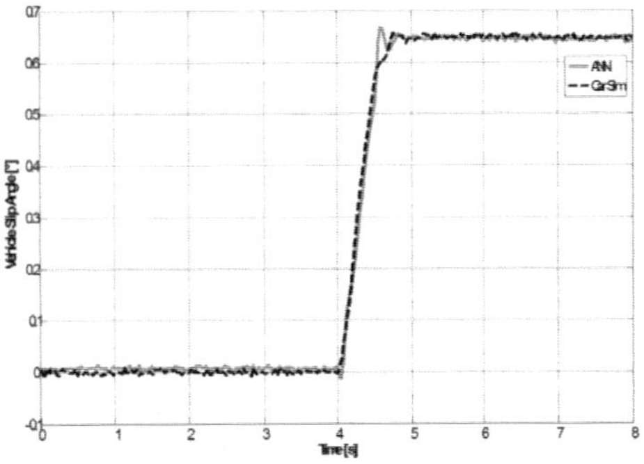

Figure 17. 45° step steer @ 40 km/h, μ=0.4.

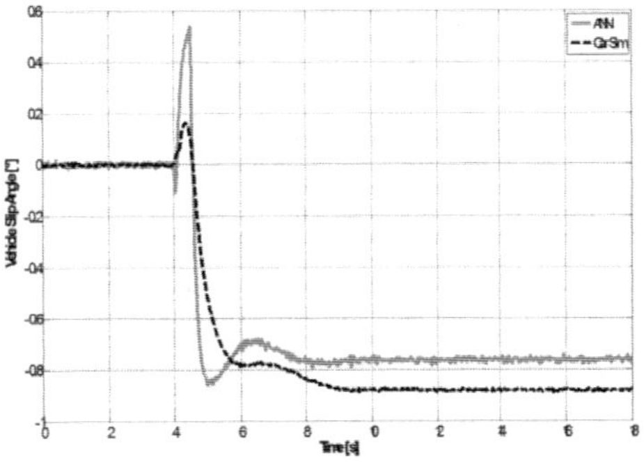

Figure 18. 45° step steer @ 50km/h, μ=0.4.

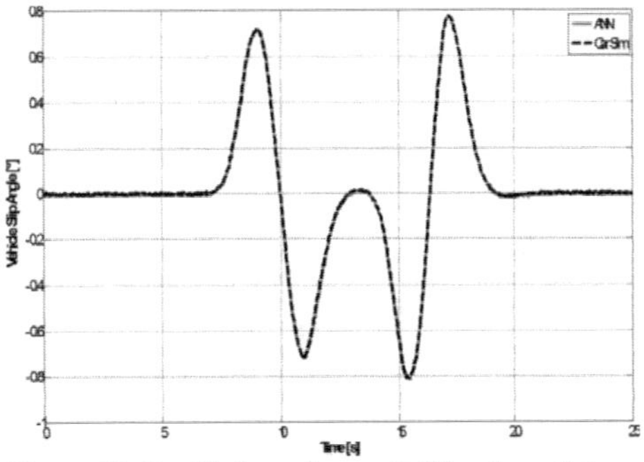

Figure 19. Double lane change @ 30km/h, μ=0.4.

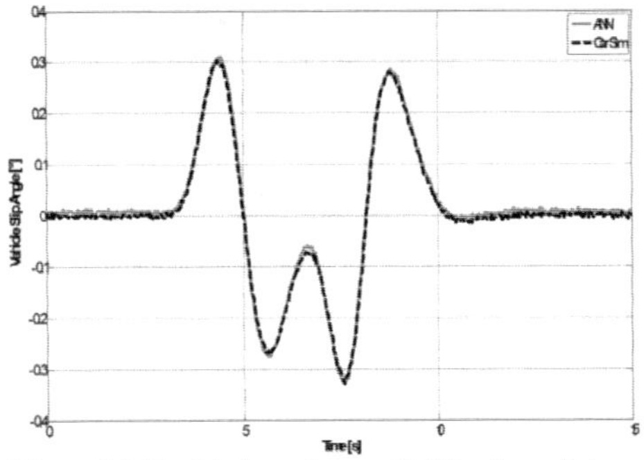

Figure 20. Double lane change @ 40km/h, μ=0.4.

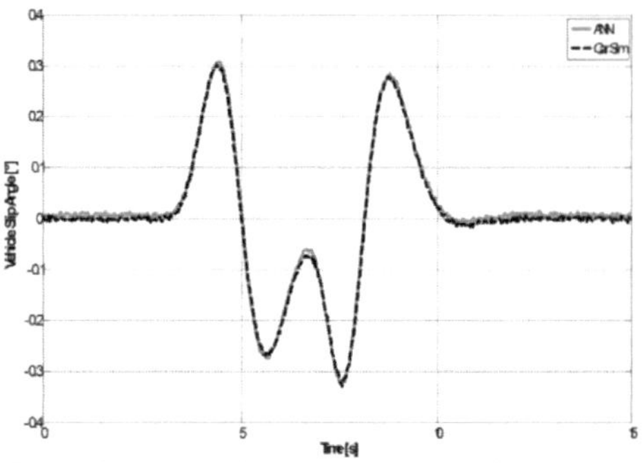

Figure 21. Double lane change @ 50km/h, μ=0.4.

Therefore, results indicate that the proposed ANN is able to accurately learn the above-mentioned training data sets and hence, well trained. Indeed, the trained ANN is able to generalize a wide range of operating conditions and the predictive ability of this network for vehicle side-slip angle is satisfactory.

Finally, for generic pattern identification the network takes 1.34e-5 s (calculated using MATLAB®), hence the proposed trained ANN is able to perform real-time calculations. Increasing the number of neurons in the hidden layer the calculation time grows exponentially without a significant improvement of the estimation.

Table 1. Nomenclature.

Abbreviation	Meaning
ESP	electronic stability program
DYC	dynamic yaw control
ABS	antilock braking system
EKF	extended Kalman filter
ANN	artificial neural network
BP	back-propagation
vx	vehicle speed
δ	steering angle
ay	vehicle lateral acceleration
ax	vehicle longitudinal acceleration
ψ	vehicle yaw rate
β	vehicle side-slip angle
MSE	mean squared error
DLC	double lane change
SS	step steer
μ	friction coefficient

4 Conclusion

In this paper, a method to estimate vehicle side-slip angle during transients and steady state conditions has been presented. The proposed technique is based on a simple artificial neural network able to perform real time estimations.

A custom design network training manoeuvre is proposed in order to keep the network as simple as possible. This approach leads to use a simple 5-10-1 layout therefore reducing the computational burden of the estimation algorithm thus making it suitable for real-time, on-line applications.

Despite the simplicity of the network, simulation results have proved the reliability and effectiveness of the proposed methodology. An accurate comparison with the estimated vehicle side-slip angle is obtained by means of the simulations performed with the CarSim® package. For the sake of completeness, results are summarized in Table 2.

This method allows the user to better evaluate the vehicle behaviour and to understand the critical points regarding stability and active safety even when the direct measurement of the side-slip angle is not available.

Table 2. Results summary.

Test	Speed [km/h]	μ	Mean Error [%]
DLC	30	1	1.75
DLC	60	1	1.92
DLC	90	1	4.85
DLC	30	0.7	1.55
DLC	60	0.7	1.63
DLC	90	0.7	4.96
DLC	30	0.4	2.12
DLC	40	0.4	2.67
DLC	50	0.4	3.32
SS	30	1	1.98
SS	60	1	1.95
SS	90	1	3.60
SS	30	0.7	1.88
SS	60	0.7	2.98
SS	90	0.7	3.58
SS	30	0.4	1.76
SS	40	0.4	2.15
SS	50	0.4	15.78

References

1. Marchesin, F.P.; Barbosa, R.S.; Alves, M.A.L.; Gadola, M.; Chindamo, D.; Benini, C. (2016). Upright mounted pushrod: the effects on racecar handling dynamics. The Dynamics of Vehicles on Roads and Tracks. Proceedings of the 24th Symposium of the International Association for Vehicle System Dynamics, IAVSD 2015. 543-552.
2. Benini, C; Gadola, M; Chindamo, D; Uberti, S; Marchesin, F.P.; Barbosa, R.S. (2017). The influence of suspension components friction on race car vertical dynamics. Vehicle System Dynamics. 55(3): 338-350.
3. Gadola, M., Chindamo, D., Romano, M., Padula, F., Development and Validation of a Kalman Filter-Based model for Vehicle Slip Angle Estimation, Vehicle System Dynamics, 2004, 52(1), 68-84.
4. H. Du, J. Lam, K.-C. Cheung W. Li, N. Zhang, Side-slip angle estimation and stability control for a vehicle with a non-linear tyre model and a varying speed, Proceedings of the Institution of Mechanical

Engineers, Part D: Journal of Automobile Engineering, 229 (2015) 486-505.

5. D.W. Pi, N. Chen, J.X. Wang, B.J. Zhang, Design and evaluation of sideslip angle observer for vehicle stability control, International Journal of Automotive Technology, 12 (2011) 391-399.
6. Chindamo, D., Economou, J.T., Gadola, M., Knowles, K., A neurofuzzy-controlled power management strategy for a series hybrid electric vehicle, Proceedings of the Institution of Mechanical Engineers, Part D: Journal of Automobile Engineering, 2014, 228(9), 1034-1050.
7. A. Lucas, M. Duran, M. Carmona, M. Lapuerta, Modeling diesel particulate emissions with neural networks, Fuel 4 (2001) 548–593.
8. M. Gwak, K. Jo, M. Sunwoo, Neural-network multiple models filter (NMM)-based position estimation system for autonomous vehicles, International Journal of Automotive Technology, 14 (2013) 265-274.
9. V. Cirovic, D. Aleksendric, D. Smiljanic, Longitudinal wheel slip control using dynamic neural networks, Mechatronics, 23 (2013) 135-146.
10. Crema, C.; Depari, A.; Flammini, A.; Vezzoli, A.; Benini, C.; Chindamo, D.; Gadola, M.; Romano, M. (2015). Smartphone-based system for vital parameters and stress conditions monitoring for non-professional racecar drivers. Proceedings of the 2015 IEEE SENSORS. 7370521.
11. M. Gadola, D. Vetturi, A. Magalini, Use of an innovative multi-input one-output neural network for experimental data numerical description, 7th International Research/Expert Conference – TMT 2003, Lloret de Mar, Barcelona, Spain, September 2003.
12. H.Sasaki, T.Nishimaki, A side-slip angle estimation using neural network for a wheeled vehicle, SAE Technical Paper 2000-01-0695, 2000.
13. Wang Wei, Bei Shaoyi, Zhang Lanchun, Zhu Kai, Wang Yongzhi, and HangWeixing, Vehicle Sideslip Angle Estimation Based on General Regression Neural Network, Mathematical Problems in Engineering, Volume 2016, Article ID 3107910, 7 pages.
14. Dal Bianco, N.; Lot, R.; Gadola, M., Minimum time optimal control simulation of a GP2 race car. Proceedings of the Institution of Mechanical Engineers, Part D: Journal of Automobile Engineering. Article in press.
15. M. Kato, K. Isoda, H. Yuasa, Estimation of vehicle slip angle with artificial neural network. SAE Technical paper, 15 (1994) 73-86.
16. S. Melzi, E. Sabbioni, On the vehicle sideslip angle estimation through neural networks. Mechanical Systems and Signal Processing, 25 (2011) 2005-2019.
17. P.D. Wesserman, Neural Computing Theory and Practice, Van Nostrand Reinhold, 1989.
18. W-J. QI, P. Zhang, Z-L. Deng, Robust Sequential Covariance Intersection Fusion Kalman Filtering over Multi-agent Sensor Networks with Measurement Delays and Uncertain Noise Variances, Acta Automatica Sinica, 40 (2014) 2632-2642.
19. A. Parlak, Y. Islamoglu, H. Yasar, A. Egrisogut, Application of Artificial Neural Network to Predict Specific Fuel Consumption and Exhaust Temperature for a Diesel Engine, Applied Thermal Engineering, 26 (2006) 824–828.
20. M.T. Hagan, H.B. Demuth, M. Beale, Neural Network Design, PWS Publishing Company, Boston, 1995.
21. Levenberg, K., A Method for the Solution of Certain Problems in Least-Squares, Quarterly Applied Math. 2, pp. 164-168, 1944.
22. Marquardt, D., "An Algorithm for Least-Squares Estimation of Nonlinear Parameters," SIAM Journal Applied Math., Vol. 11, pp. 431-441, 1963.
23. M. Aydinalp, V.I. Ugursal, A.S. Fung, Predicting residential appliance, lighting, and space cooling energy consumption using neural networks. The Fourth International Thermal Energy Congress, Cesme, Turkey, July.
24. S. Haykin, Neural Networks, A comprehensive foundation, McMillian College Publishing Company, New York, 1994.

Braking Control for Improving Ride Comfort

Jonghyup Lee[1], Seibum Choi[1]

[1,1a]*KAIST, Department of Mechanical Engineering, 34141 Daejeon, Republic of Korea*

Abstract. While many vehicle control systems focus on vehicle safety and vehicle performance at high speeds, most driving conditions are very low risk situations. In such a driving situation, the ride comfort of the vehicle is the most important performance index of the vehicle. Electro mechanical brake (EMB) and other brake-by-wire (BBW) systems have been actively researched. As a result, braking actuators in vehicles are more freely controllable, and research on improving ride comfort is also possible. In this study, we develop a control algorithm that dramatically improves ride comfort in low risk braking situations. A method for minimizing the inconvenience of a passenger due to a suddenly changing acceleration at the moment when the vehicle is stopped is presented. For this purpose, an acceleration trajectory is generated that minimizes the discomfort index defined by the change in acceleration, jerk. A controller is also designed to track this trajectory. The algorithm that updates the trajectory is designed considering the error due to the phase lag occurring in the controller and the plant. In order to verify the performance of this controller, simulation verification is completed using a car simulator, Carsim. As a result, it is confirmed that the ride comfort is dramatically improved.

1 Introduction

Most of the control researches on braking and turning of the vehicle are carried out to improve the safety or the performance of the vehicle in a dangerous situation. However, most of the driving situation is a non - dangerous driving situation which does not require the extreme performance of the vehicle. As the autonomous driving researches actively, it is expected that the performance need for the extreme situation of the vehicle is further reduced. For such non- dangerous driving situations, the ride quality of the vehicle is the most important performance indicator. In fact, studies on the relief of passengers' ride comfort have been actively conducted for vehicles with a significantly lower probability of accidents such as wheelchair, elevator [1],[2]. However, in the case of the conventional automobile, the autonomous control cannot be performed due to the hardware structure which depends on the driver input. As the brake by wire (BBW) system such as electro mechanical brake system (EMB) and electro wedge brake system (EWB) is developed in the recent vehicle system, the vehicle can freely control the behavior of the vehicle itself in addition to the cylinder input of the driver. This allows the control system to get rid of the constraints on multiple control situations that depend only on the existing operator input, so that braking control for ride comfort is also possible in the vehicle.

The uncomfortable ride is caused by amount of acceleration and jerk, which is the amount of change in acceleration. Because of the acceleration of the vehicle, the passenger receives the inertia force and the body receives the force. The greater the magnitude of the force or the greater the change in force, the more the passenger feels discomfort. Therefore, in this study, control is performed for the longitudinal braking situation to increase the ride comfort. Define the discomfort index for the ride comfort and create a desired acceleration trajectory that minimizes the discomfort index. In addition, an acceleration controller capable of keeping up with this is designed to perform braking control for ride comfort.

The composition of this paper is as follows. A detailed description of the control target in this study is presented in section 2. Controller design to meet the control target is done in section 3. Then, the simulation verification and control results via the Carsim are shown in section 4. Finally, conclusions are given in Section 5.

2 Control target

The moment when a passenger feels uncomfortable during braking is roughly classified into three types. The first is the inconvenience that the body suddenly slips forward when braking starts, the second is the inconvenience of enduring the inertia force generated in front of the braking system, and the inconvenience that the body suddenly leans back when the vehicle finally stops. This inconvenience is caused by the acceleration that necessarily occurs for the vehicle to stop, and it is determined by the input of the driver for the deceleration capability which is entirely determined by the driver. In

 https://doi.org/10.1051/matecconf/201816602002

other words, if the acceleration and the jerk are reduced in size to improve the ride comfort, the ride quality is improved but the possibility of an accident is increased due to the increase of the braking distance. In particular, since the starting point of braking and the amount of deceleration during braking greatly influence the braking distance, it is not suitable as a target of braking control for ride comfort. However, in the case of jerk which occurs at the moment when the vehicle is stopped as shown in Fig. 1, since the control is performed at a low speed, the braking distance is not greatly affected and the ride quality can be improved.

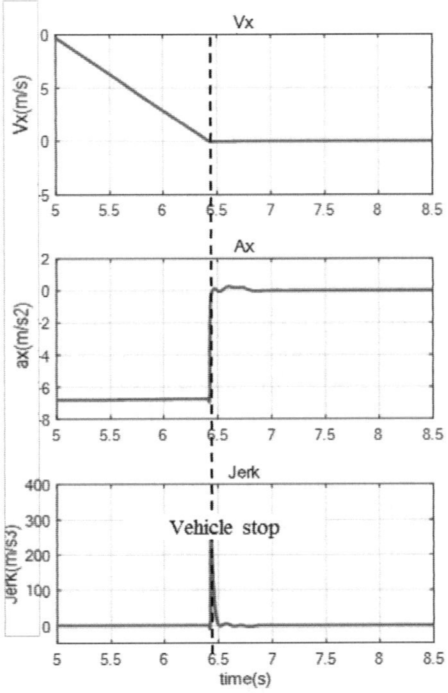

Figure 1. Vehicle speed, acceleration and jerk at vehicle stop.

The principle of the jerk that occurs when the vehicle stops is as follows. Even if the vehicle has a small deceleration while stopping, the acceleration suddenly disappears to zero because the relative motion with the ground disappears as soon as the vehicle stops. As a result, the magnitude of the acceleration of the vehicle is greatly reduced discontinuously, and it is represented by a very large jerk. The passenger suddenly feels the uncomfortable feeling of suddenly turning back due to the sudden change of acceleration. In this study, the magnitude of the suddenly changing acceleration is slowly decreased from immediately before stopping, so that the instantaneous acceleration of the vehicle can be converged to zero. If control is successful and successful behavior is achieved, it can be seen that acceleration and velocity converge smoothly to zero at the same time as shown in Fig. 2, and jerk is also considerably reduced in size

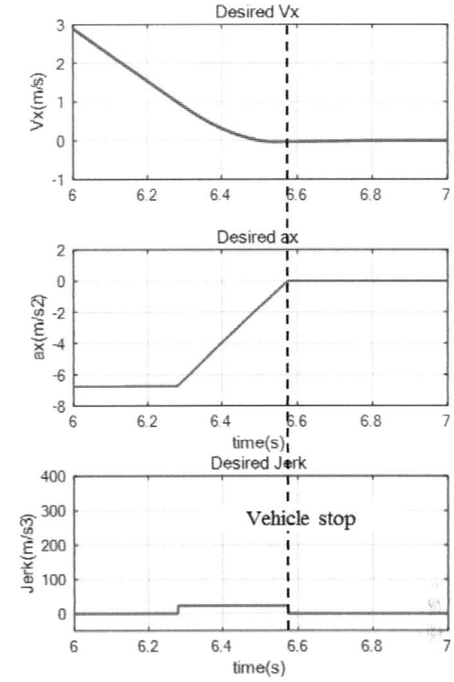

Figure 2. Desired vehicle speed, acceleration and jerk at vehicle stop for ride comfort

3 Controller design

The structure of the controller designed in this study is divided into the part that generates the desired acceleration trajectory for the control target and the controller that can track it well.

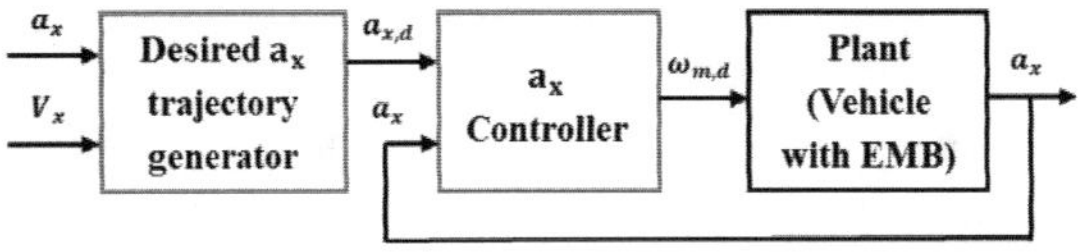

Figure 3. Control algorithm overview

3.1. Desired acceleration trajectory generator

The minimum jerk model proposed by Flash and Hogan for the objective evaluation of ride comfort has been studied to minimize this value by integrating the mean of the jerk squares as shown in Eq. (1) [3].

$$C = \int_{t_i}^{t_f} \left(\frac{d^3 x}{dt^3} \right)^2 dt \tag{1}$$

Where d^3x/dt^3 is longitudinal jerk, t_i is initial time of the control, and t_f is final time of the control. In order to minimize the discomfort index in the braking situation, the magnitude of acceleration at the moment when the vehicle is stopped should also converge to 0 at the same time in order to prevent the acceleration of the moment when the vehicle stops. In order to minimize the discomfort index under these conditions, it should be constant because the value obtained by integrating the jerk should be equal to the magnitude of the acceleration at the moment of initiation of control as shown in equation (3).

$$a_x(t_f) = 0,\ v_x(t_f) = 0 \tag{2}$$

$$\int_{t_i}^{t_f} \frac{d^3x}{dt^3} dt = a_x(t_f) - a_x(t_i) = -a_x(t_i) \tag{3}$$

$$j_{x,d}(t) = \frac{d^3x_d}{dt^3} = const. \tag{4}$$

$$a_x(t_f) = a_x(t_i) + j_{x,d}\cdot(t_f - t_i) = 0 \tag{5}$$

$$v_x(t_f) = v_x(t_i) + a_x(t_i)\cdot(t_f - t_i) + \frac{1}{2} j_{x,d}\cdot(t_f - t_i)^2 = 0 \tag{6}$$

$$j_{x,d} = \frac{a_x(t_i)^2}{2v_x(t_i)} \tag{6}$$

Equations (4) and (5) express the condition that the speed and the acceleration become 0 at the same time when the vehicle stops. Through these equations, the desired jerk is expressed as (5), and the desired acceleration trajectory generator generates the desired acceleration trajectory through the acceleration and velocity information at the start of braking control.

$$a_x(t) = a_x(t_i) + j_{x,d}\cdot(t - t_i) \tag{7}$$

If the lower controller produces an ideal performance and follows exactly the designed acceleration trajectory, the discomfort index C is significantly improved as compared with the control without the control as shown in Table 1.

Table 1. Discomfort index when ideal braking control is performed

	Discomfort index
w/o control	3254
Ideal control	153

3.2. Plant modelling and acceleration controller

In this study, braking control was performed through Electro Mechanical Brake system (EMB). It is a system that applies force to the brake through a motor, not a brake that moves the piston through the existing hydraulic pressure. A typical DC motor is used for the motor, and the dynamics inside the EMB are expressed simply as Equation 8-10 [4].

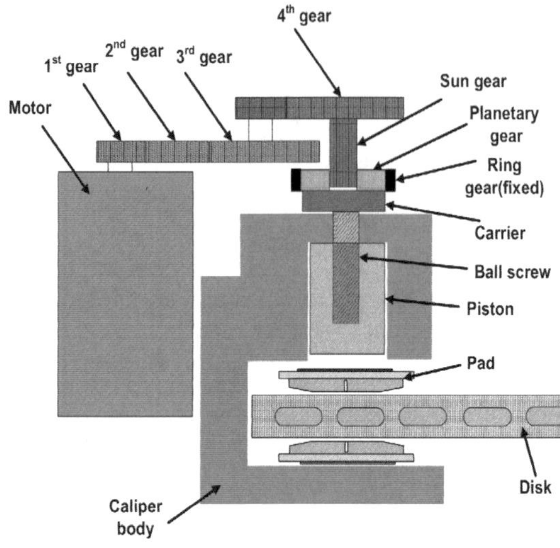

Figure 4. Electro mechanical brake system

$$T_m = k_t i_q \tag{8}$$

$$J\dot{\omega} = T_m - T_f - T_L \tag{9}$$

$$F_{cl} = k\theta \tag{10}$$

To control the clamping force of the EMB with this dynamics, the desired motor speed is calculated through a longitudinal acceleration controller, which is implemented via PI type speed controller and current controller.

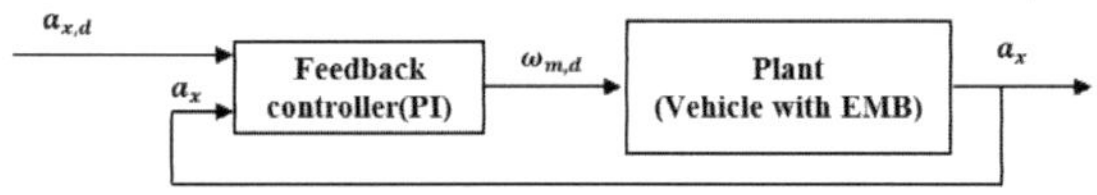

Figure 5. Internal control algorithm of EMB

Since the acceleration of the vehicle is proportional to the motor position of the EMB, the motor position control of the EMB is implemented through the PI controller to track the desired acceleration trajectory generated in Section 3.1.

3.3. Trajectory update algorithm

The target acceleration trajectory in this study should have zero longitudinal velocity and longitudinal acceleration simultaneously. Therefore, if perfect tracking is not achieved, it is difficult to achieve the existing goal. Although the EMB has a wide band width due to its speed in terms of reaction speed compared to conventional hydraulic brakes, control may fail even with a slight phase lag due to the controller and the plant. Therefore, it is necessary to design an algorithm that increases the control success rate by continuously updating the desired trajectory while controlling the desired trajectory rather than tracking it at a time when control starts.

In this study, we assume that the sampling rate of the speed measured by GPS is 20Hz, and update the trajectory every 50ms to implement tracking control.

4 Simulation Results

The simulation algorithm was implemented through Matlab Simulink and validated through a vehicle simulator, Carsim. The scenario started from the moment when the vehicle was at braking and the vehicle speed was 1 m / s.

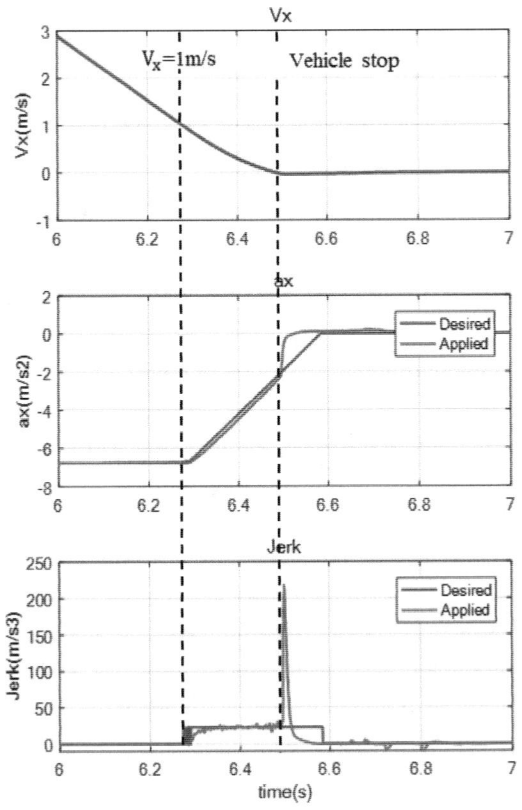

Figure 6. Control result without update algorithm

Figure 6 shows the result of tracking control after creating one desired trajectory. The jerk was decreased compared with the case where the control was not performed. However, as the phase lag caused the vehicle to stop, there was an acceleration, which resulted in a large jerk.

Figure 7 shows the result of simulation by adding the algorithm that updates the desired trajectory. When trajectory is updated every 50 ms, it is confirmed that a desired trajectory is corrected to compensate for the error caused by the initial phase lag of the control. Thus, it is confirmed that the control finally succeeded.
The control performance for each algorithm is summarized in Table 2.

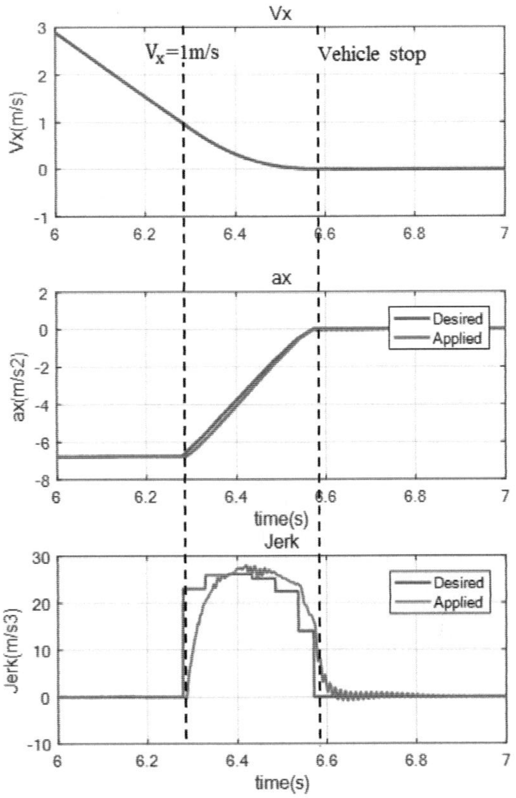

Figure 7. Control result with update algorithm

Table 2. Control performance comparison

	Discomfort index	Max jerk
w/o control	3254	718
Ideal control	153	23
Without update algorithm	423	217
With update algorithm	175	31

The discomfort index is much smaller than without control when there is no update algorithm, but it is significantly lower than expected performance. When the trajectory update algorithm is added, the result is similar to the expected performance in the ideal control situation. The biggest side effect of braking control for ride comfort is the increase of braking distance. However, since the control starts at a low speed of 1m / s just before the vehicle stops, the braking distance increased by 2cm.

4 Conclusion

In order to control the passengers' ride comfort, we constructed an algorithm that reduces the brake input just before the vehicle stops. To do this, we created an acceleration trajectory that minimizes the discomfort index defined by the magnitude of jerk. We also designed PI controller based acceleration control algorithm and EMB internal control algorithm. We also added a trajectory update algorithm to solve the problem caused by phase lag between controller and plant. As a result, the results were significantly better than those without control. Although there was a great improvement in ride comfort, the increase in the braking distance, which is a side effect of this control, was very small, and the actual implementation feasibility of the algorithm is also sufficient.

Acknowledgment

This work was supported in parts by the National Research Foundation of Korea grant funded by the Korea government (MSIP) (No.2017R1A2B4004116), the Ministry of Science and ICT, Korea, under the Information Technology Research Center support program (IITP-2017-2012-0-00628) supervised by the Institute for Information and communications Technology Promotion, and the BK21 Plus Program.

Reference

1. H Seki, K Ishihara, S Tadakuma, " Novel Regenerative Braking Control of Electric Power-Assisted Wheelchair for Safety Downhill Road Driving", *IEEE Transactions on Industrial Electronics*, Volume: 56, Issue: 5, May 2009
2. Jun-Koo Kang and Seung-Ki Sul, "Vertical-vibration control of elevator using estimated car acceleration feedback compensation," in *IEEE Transactions on Industrial Electronics*, vol. 47, no. 1, pp. 91-99, Feb 2000
3. T Flash, N Hogan, "The coordination of arm movements: an experimentally confirmed mathematical model", *Journal of Neuroscience*, 1 July 1985, 5 (7) 1688-1703
4. C Line, C Manzie, MC Good, J Nell, P Rieth, "Electromechanical Brake Modeling and control", *IEEE Transactions on Control Systems Technology*, Volume: 16, Issue: 3, May 2008

A Braked Tail Cover Separation Concept and Simulation of Separation Process

Mei Huang, Yi Jiang and Boman Wang

School of Aerospace, Beijing Institute of Technology, Beijing, China

Abstract. Large-scale missiles are generally catapulted by the ejector. And a tail cover is used to protect the first stage rocket engine. When the missile reaches the predetermined height, the tail cover should be separated timely so that the engine can work properly. In this paper, a braked tail cover separation concept is proposed. And a new type of tail cover separation device is designed according to the concept. Firstly, the forces analysis and the internal ballistic simulation of ejection are conducted before setting working conditions. And then, the finite element software Abaqus is adopted to achieve the simulation of tail cover separation process. Through analyzing the obtained results, it's concluded that the braked tail cover separation concept is feasible. Moreover, the influence of material changes on the braking effects is also studied.

1 Introduction

At present, large-scale cylindrical missile generally use cold launch to complete the launch process [1]. That is, the high temperature and high pressure gas produced by the gas generator at the bottom of the launcher pushes the missile out of the tube. when the missile reaches the predetermined height, the rocket engine ignites and the missile starts the flight. In order to avoid the damage the gas caused to the equipment and the engine, a tail cover is necessary to protect the missile in the process of ejection. when the ejection is finished, the tail cover should be separated timely so that the engine can work properly. Moreover, the tail cover should be deviated from the launcher to avoid smashing the personnel and equipment [2-3]. Therefore, the tail cover separation device should meet the following requirements: fast, reliable and safe [4].

According to the requirements above, a new tail cover separation scheme is present in this paper and a new type of tail cover separation device is designed.

2 Tail cover separation concept and tail cover separation device

2.1 Tail cover separation concept

Firstly, the braked tail cover separation concept is proposed. Its schematic diagram is shown in Fig. 1. The projectile and the tail cover are connected by a locating pin. When the missile is ejected, the tail cover is arrested in the tube by the tail cover separation device. For the projectile is not screwed with the tail cover, it flight at a natural speed. In this scheme, the tail cover separation device is a vital technology.

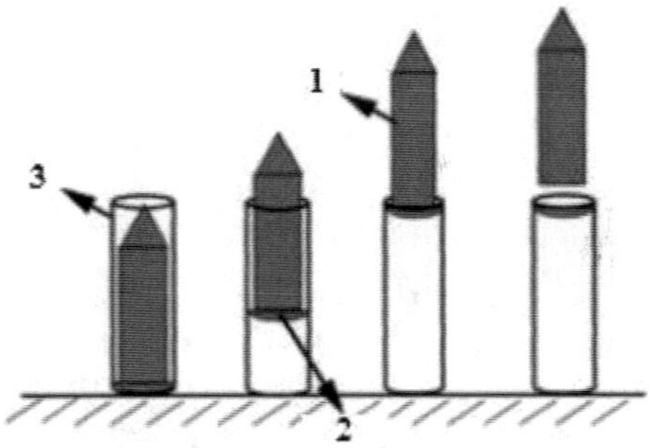

Figure 1. Schematic diagram of the braked tail cover separation concept: 1-projectile; 2-tail cover; 3-launch tube.

2.2 Tail cover separation device

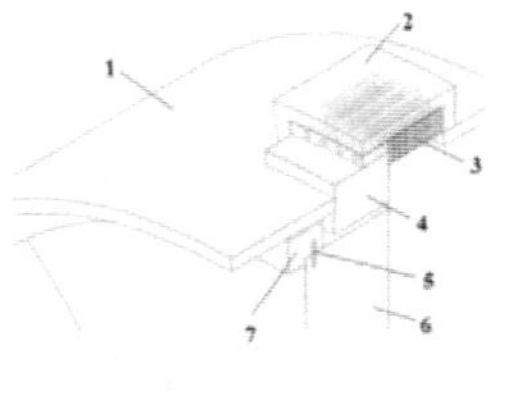

Figure 2. Structure of the braked tail cover separation device : 1-launch tube; 2-inner wall; 3-metal slice combination; 4-trasmission device; 5-connecting pin; 6-projectile; 7-tail cover.

Based on the braked tail cover separation concept, a new type of tail cover separation device is designed. It is shown in Fig. 2.

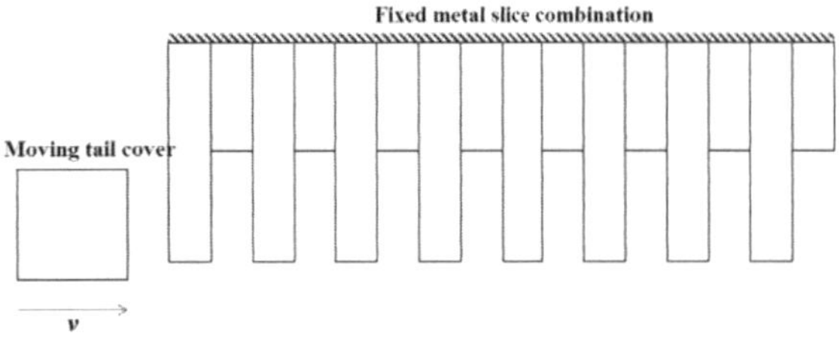

Figure 3. The basic working principle diagram of the device

Fig. 3 shows the basic working principle of the device. when the tail cover with great kinetic energy moves to the mouth of the launch tube, it is arrested by the metal slice combination, and the metal slice combination is cut to realize the deceleration of the tail cover.

In the way of 90° ×4, the device is installed at the mouth of the launch tube. The projectile and the tail cover are connected by locating pins. And the two rush to the braked device with great kinetic energy. After the collision with the transmission device, the positioning pin bends first. Then the missile and the tail cover start separation and the tail cover rush to the metal slice combination fixed on the braked device. When contacting with the moving tail cover, the first metal slice is sheared and deformed, and then the metal combination is sheared and deformation one after another until the tail cover decelerates to zero. At the same time, the missile flies out of the launcher. Now, the separation of the tail cover and the projectile has been achieved.

3 Calculation and simulation

Braked device is a key component in the tail cover separation scheme, which needs to withstand the force and thermal environment in the process of ejection. Its braking effects are to studied through numerical simulation. Before the simulation, some calculations are necessary. First, according to the results of internal ballistic simulation during the ejection, the separation process of the tail device is calculated and analyzed. Then, the numerical simulation model is established in Abaqus. Next, set the parameters in conjunction with the calculated results in Abaqus. At last, the simulation is run and the obtained results are processed and analyzed.

3.1. Calculation

3.1.1 Calculation condition

Table 1. Initial design parameters .

Parameters	value
Mass of the tail cover	$80kg$
Radius of the tail cover	$1.49m$
Velocity of the tail cover	$38m/s$
Stroke of ejection	$2m$

Before the simulation is conducted, some calculation is necessary.the initial design paramerers are shown in Table 1.including m the mass of tail cover, r radius of tail cover, v the velocity of tail cover when it hits the device and s the stroke of ejection.

3.1.2 Forces analysis

When the tail cover rushes to the launcher mouth, some forces act on it. including G the gravity, F_r the gas pressure in the launcher tube , F_s the braking force of the braked device , F_a the external atmospheric pressure and F_g the inertia force of the tail cover, ignoring the friction between the sealing ring of tail cover and the tube wall, the pin hole friction, and other smaller forces. Forces analysis is shown in Fig. 4.

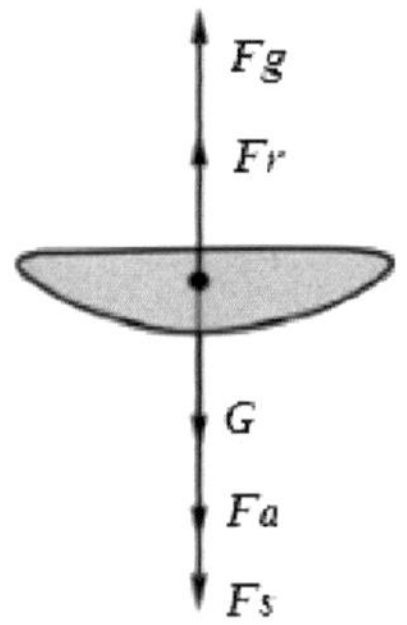

Figure 4. The forces analysis of the tail cover separation device

Where, the gravity of the tail cover is shown in (1).

$$G = mg = 784N \tag{1}$$

And the external atmospheric pressure is shown in (2)

$$F_a = P_o \cdot \pi r^2 = 176543N \tag{2}$$

3.1.3 The initial forces

According to the calculation conditions, the velocity of the tail cover decelerates from 38 m / s down to 0 m / s in the 2m stroke, assuming the it decreases evenly, it can be conclude in (3):

$$F_g = \frac{mv^2}{2s} = 28880N \tag{3}$$

3.1.4 the maximum gas pressure in the launch tube

Take the calculation conditions above into internal ballistic program, the curve of internal ballistic simulation during the ejection is shown in Fig. 5.

During the ejection process, the pressure of launch tube increases rapidly at first and then drops slowly.

From Figure5, it can be seen that the final pressure when the tail cover hits the braked device is about 0.5MPa (absolute pressure). So we can get the value of F_r.

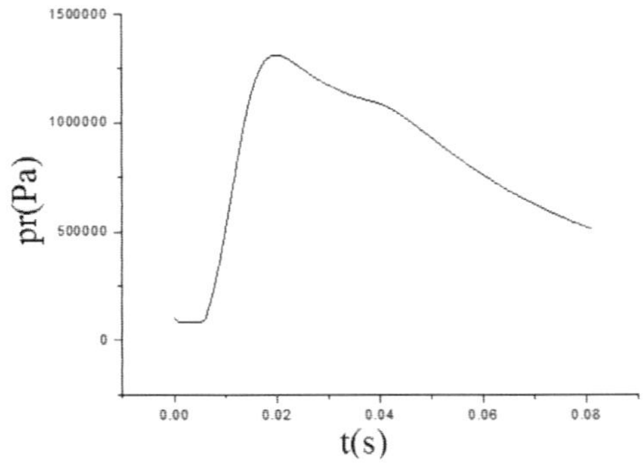

Figure 5. Cure of the gas pressure P_r in launch tube.

$$F_r = P_r \cdot \pi r^2 = 871389N \quad (4)$$

3.1.5 The braking force

According to the force balance conditions, the braking force is calculated in (5):

$$F_s = F_g + F_r - G - F_a = 722942N \quad (5)$$

3.2 Working condition setting

Based on the working principles and calculation conditions above, the numerical simulation model is established in Abaqus. The braking effect of the tail cover is studied. Moreover, the influence of material change on the braking effect is also studied.

In the structure of the tail cover separation device, the metal slice combination is made of aluminum while the other parts are made of steel. Thus, three working conditions have been designed by changing the material property of different structures.

Table 2. Material property of working condition 1.

Material	Aluminum	Steel
Density(kg/m3)	2.8×10^3	7.87×10^3
Young's modulus(MPa)	72400	210000
Poisson's ratio	0.33	0.3
Yield stress(MPa)	395	1220

Compared to the working condition 1 shown in Table 2, working condition 2 increases the yield stress of the aluminum. And the material parameters of the working condition 2 are shown in Table 3.

Table 3. Material property of working condition 2.

Material	Aluminum	Steel
Density(kg/m3)	2.8×10^3	7.87×10^3
Young's modulus(MPa)	72400	210000
Poisson's ratio	0.33	0.3
Yield stress(MPa)	758.291	1220

Compared to the working condition 1, working condition 3 reduces the yield stress of the steel. The material parameters are shown in Table 3.

Table 4. Material property of working condition 3.

Material	Aluminum	Steel
Density(kg/m3)	2.8×10^3	7.87×10^3
Young's modulus(MPa)	72400	210000
Poisson's ratio	0.33	0.3
Yield stress(MPa)	395	785

Compared to the working condition 1, working condition 3 changes the yield stress of the steel.

3.3 Simulaton and results analysis

3.3.1 Simulation

Three dimensional model of the tail cover separation device is established in Creo, which is imported into Abaqus. Then we give the material properties, apply boundary constraints and loads, mesh the model and submit file for calculation. Abaqus software is uesd to simulate the process of tail cover separation by using finite element method.

3.3.2 Results and discussion

The obtained simulation results are compared under different working conditions . At first, a point in the tail cover is selected to measure the velocity, the braking distance, and the whole kinetic energy of the tail cover. Besides, the deformation of the device is also measured. Curves of different variable in the three working condition are shown in Fig. 7 to Fig. 10.

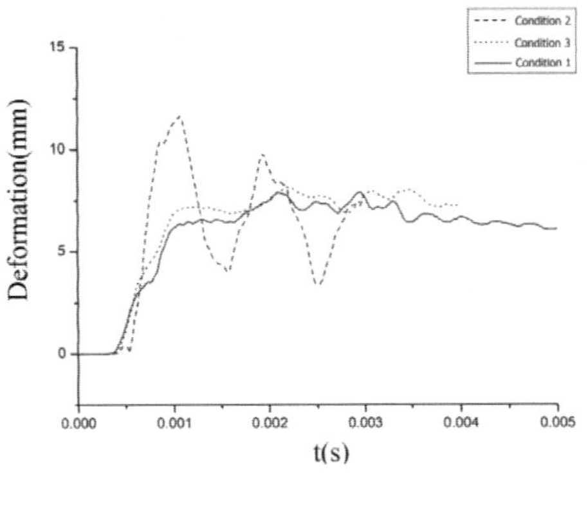

Figure 6. Deformation of the device in the three conditions.

Fig. 7 shows that the trend of structural deformation of the working condition 1 and the working condition 3 is almost the same and they are relatively stable after a sharp rise. The curve of the working condition 2 fluctuates violently.

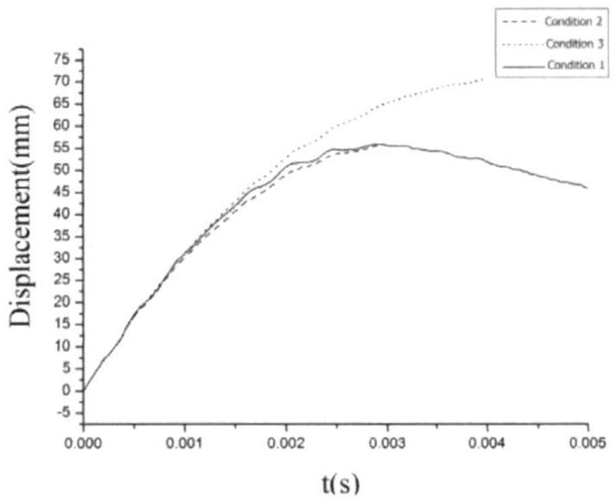

Figure 7. Braking distance in the three conditions.

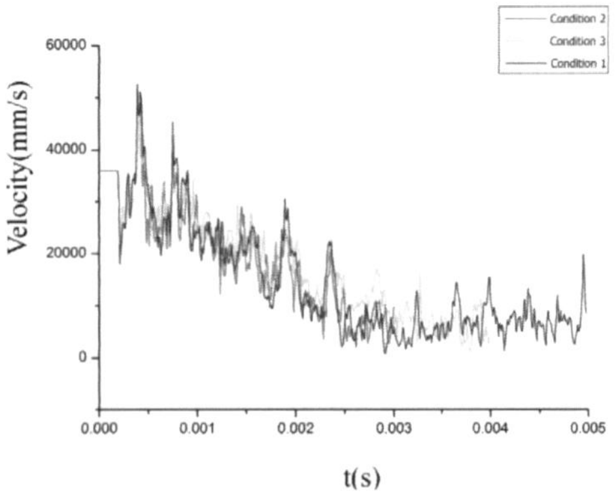

Figure 8. Velocity of the device in the three conditions.

Fig. 8 shows that the curves of braking distance in the working condition 1 and 2 almost coincide, and the maximum amplitude is 55mm. While the curve in the working condition 3 is increasing with time and its maximum is 68mm.

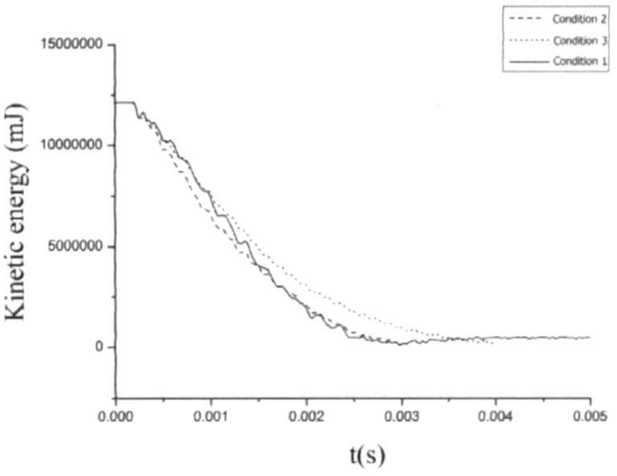

Figure 9. Kinetic energy in the three conditions.

From the Fig. 10, we can see that the trend of the whole kinetic energy in the working condition 1 is closer to that in the working condition 3, and then the difference is larger. For the condition 2, there is a big difference compared to the first two in the beginning. In the later stages, the two trends have become closer. After a while, the overall kinetic energy of the three tend to zero.

We define that the tail cover separation has finished when the velocity of the tail cover tends to 0. From figures above, it can be seen that the braking time of the three working conditions is 0.05s, 0.03s, 0.04s respectively.

Considering that the metal slice combination is made of aluminum while the other parts are made of steel, the influence of material change on the braking effect of the device can be summarized in Table 5.

Table 3. the influence of material change on the braking effect.

Braking effects Material change	**Increasing yield stress of the shear slice combination**	**Increasing yield stress of other parts of the device**
deformation	greater	not significant
braking distance	not significant	smaller
deceleration effects	tendency is not significant	tendency is not significant
the absorbing capacity of kinetic energy	Absorb faster during high-speed stage	Absorb faster during low-speed stage

4 Conclusion

In this paper, a braked tail cover separation concept is proposed. Based on this, a new type of tail cover separation device is designed. The numerical simulation model is established in Abaqus. By comparing different working conditions, the braking effects of this device are studied. The obtained results show that the braked tail cover separation scheme is feasible. And the device can separate the tail cover from the projectile quickly and reliably. Furthermore, we can enhance the braking effects by increasing yield stress of the shear slice combination or lowing yield stress of other parts of the device appropriately.

References

1. R. Chao, R. Yang, X.Q. Zhao, J. Xie. An arrested tail cover separation technology. Missiles and Space Vehicles **3**, 12-14(2013)
2. N. Hua, J. Yan. A new tail cover separation concept and simulation of separation process. Missiles and Space Vehicles **2**,4-8(2005)
3. R.Y. Jia, T. Wang, Z.Y. Jiang, F. Hu, W.H. Zhang. Uncertainty analysis of the rocket tail cover separation. Journal of National University of Defense Technology **6**,88-92(2014)
4. H. Wang, X.C. You, S. Lin. Numerical simulation of rocket aft dome separation. Missiles and Space Vehicles **5**, 15-19(2011)

A Novel Adaptive Particle Swarm Optimization Algorithm with Foraging Behavior in Optimization Design

Yan Liu[1] , Qi Wang[2] , Guoyi He[3], Li Zhang[1] and Jiao Wang[1]

[1] *Nanchang Hangkong University School of Aircraft Engineering,student,2017, China*
[2] *Nanchang Hangkong University School of Aircraft Engineering, professor,2017, China*
[3] *Nanchang Hangkong University School of Aircraft Engineering, associate professor,2017, China*

Abstract. The method of repeated trial and proofreading is generally used to the convention reducer design, but these methods is low efficiency and the size of the reducer is often large. Aiming the problems, this paper presents an adaptive particle swarm optimization algorithm with foraging behavior, in this method, the bacterial foraging process is introduced into the adaptive particle swarm optimization algorithm, which can provide the function of particle chemotaxis, swarming, reproduction, elimination and dispersal, to improve the ability of local search and avoid premature behavior. By test verification through typical function and the application of the optimization design in the structure of the reducer with discrete and continuous variables, the results are shown that the new algorithm has the advantages of good reliability, strong searching ability and high accuracy. It can be used in engineering design, and has a strong applicability.

1 Introduction

The gear reducer is the transmission between the prime mover and the working machine which can reduce the speed and increase the torque. It is widely used in industry, mining enterprises, transportation, construction and other departments of the mechanical parts[1].

The traditional design method of cylindrical gear reducer is used to compare existing reducers and develop an initial design, which is according to a variety of information, literature and combined with the design experience, then the program will be checked, if the check results are passed, the program can be determined ,otherwise, the reducer will be redesigned. In order to reduce the cost of the reducer and improve design efficiency, optimized design of cylindrical gear reducer is imperative[2]

Particle swarm optimization algorithm is a kind of stochastic optimization technique based on population, the basic idea is inspired by the behavior of birds with rapid and premature in the early stage. The other algorithm is the bacterial foraging, it is based on the foraging behavior model of Escherichia coli, which is not sensitive to the initial value and the parameter selection, and the local search ability is strong[3]. However, to combinate two algorithms can not only improve the optimization speed of complex problems, but also can solve the problem of premature convergence in engineer application.

2 Algorithm and improvement

Considering the advantages of the particle swarm optimization algorithm (PSO) and the bacterial foraging algorithm (BFA), the first is early rapidity and precocious puberty[4], and the second is strong local search ability[5]. So the combination of two algorithms can not only improve the optimization speed, but also solve the premature convergence of the complex problems.

This paper presents the particle swarm optimization algorithm based on bacterial foraging behavior(BFPSO). In this method, the bacterial foraging process is introduced into the adaptive particle swarm optimization algorithm(APSO) [6], which can provide the function of particle chemotaxis, swarming, reproduction, elimination and dispersal, improve the ability of local search and avoid premature behavior. The flow chart of the proposed algorithm is as figure 1.

3 Function test

In order to verify the optimization performance of BFPSO, we use 5 classical test functions to compare BFPSO with the simple PSO and APSO. Test times is 200.

The test functions are: f1 Rosenbrock function, f2 Rastrigrin function, f3 Shubert function, f4 Ackely function, and f5 Griewank function[5]. The basic features of these test functions are shown in table 1.

In the MATLAB simulation calculation, respectively, the number of 5 test lines for the 200 calculation, the reliability of the test results shown in table 2, the results of the convergence process are compared in figure 2.

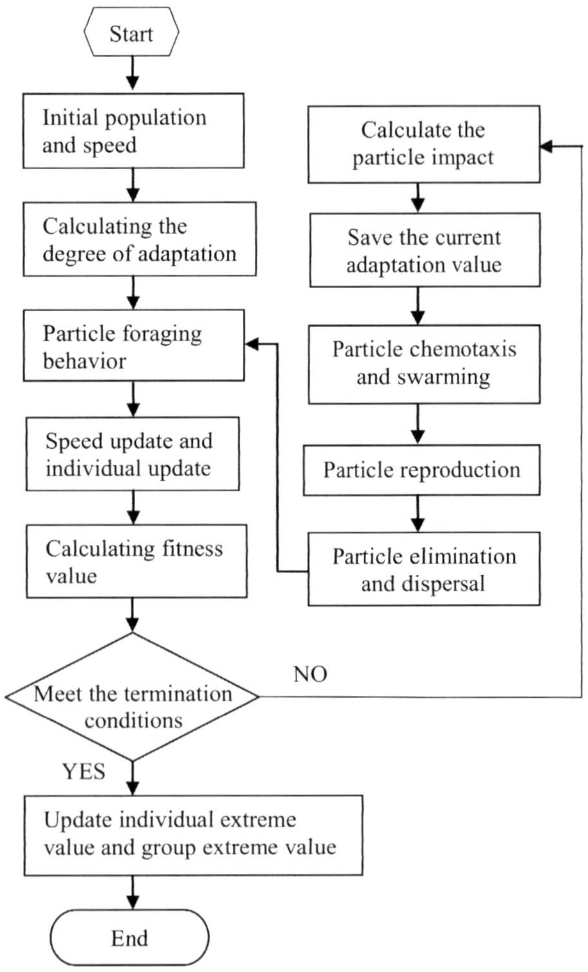

Figure 1. Flow chart of the proposed algorithm

The 200 reliability tests are shown in table 2, the accuracy of the improved algorithm is higher than the other two basic algorithms, and the accuracy requirements of the number is more, then the reliability and accuracy of the new algorithm are improved.

Compare to the convergence speed of the test functions f1,f2,f3, the new algorithm is significantly improved, the robustness is better.

For the test function f4, the convergence speed of the three algorithms is similar, because the f4 belongs to a single peak function and the extremum is at one point. The three algorithms can easily approch the extreme point, so the convergence speed is similar.

For the test function f5, the new algorithm converges slower than PSO and APSO, because the function f5 is uniform multi-peaked in the entire range of values, and the difference between the maximum absolute value point and the other extreme points is not obvious. When the PSO and APSO enter the local optimum, the BFPSO is still searching for the global optimal value. Therefore, the convergence speed of the BFPSO is slower than that of the PSO and the APSO, which shows that the global optimization capability of the BFPSO is stronger.

4 Optimization design of reducer

4.1 Design demand

This paper using the two-grade helical cylindrical gear reducer from references [7] as a analysis example. The design is to make sure the center distance is minimum in the require of bearing the design force.

Known conditions are: the input power of the high speed axle is R=6.2Kw, maximum rotate speed is n1=1450r/min, overall ratio is iΣ =31.5, the coefficient of facewidth of the gearwheel is φa=0.4, big wheel gear using No.45 steel, normalizing treatment, rigidity is ranging in(187~207)HBS, small wheel gear using No.45 steel, thermal refining rigidity is ranging in (228~255)HBS, total working time is designed not less than 10 years [7-9] .

4.2 Mathematical model

Using designing variable expression the center distance, the objective function is in Eq. (1).

$$f(x)=[x_1x_3(1+x_5)+x_2x_4(1+31.5/x_5)]/(2\cos x_6) \quad (1)$$

The constraint conditions from references [7] shown as in Eq. (2-8).

$$2<x_1<5,\quad 3.5<x_2<6,\quad 14<x_3<22 \quad (2)$$

$$16<x_4<22,\quad 5.8<x_5<7,\quad 8<x_6<15 \quad (3)$$

$$g(x)=\cos^3 x_6-3.079\times10^{-6}x_1^3x_3^3x_5\le 0 \quad (4)$$

$$g(x)=x_5^2\cos^3 x_6-1.701\times10^{-4}x_2^3x_4^3\le 0 \quad (5)$$

$$g(x)=\cos^2 x_6-9.939\times10^{-5}(1+x_5)x_1^3x_3^2\le 0 \quad (6)$$

$$g(x)=x_5^2\cos^2 x_6-1.706\times10^{-4}(31.5+x_5)x_2^3x_4^2\le 0 \quad (7)$$

$$g(x)=x_5[2(x_1+50)\cos x_6+x_1x_3x_5]-x_2x_4(x_5+31.5)\le 0 \quad (8)$$

Where x_1,x_2 is the normal module of the high speed wheel gear and the low speed wheel gear, x_3,x_4 is the number of teeth of the high speed level wheel gear and the low speed level wheel gear, x_5 is the gear ratio of the high speed level, x_6 is the helical angle of the wheel gear.

4.3 Simulation

The operating parameters of BFPSO are: the initial number of particles m=80, the learning factor $c_1=2$, $c_2=2$, the Inertia weight w=0.729, the particle chemotaxis and swarming times $N_c=50$, the reproduction times $N_{re}=4$, the elimination and dispersal times $N_{ed}=2$, the elimination and dispersal probability $P_{ed}=0.25$.

In order to verify the effectiveness of BAPSO in the structural optimization design of reducer. Takes 10 times calculations and chooses the result ($m_{n1}=2$, $m_{n2}=4$, $z_1=17$, $z_3=16$, $i_1=5.9$, $\beta=8.6334^{\circ}$, $i_2=5.339$, $z_2=100$, $z_4=85$, center distance is 323.8160,then corrected the center distance to 324, the helix angle is adjusted to $\beta=10.0788^{\circ}$) which is the biggest center distance to compare with the

Table 1. Basic characteristics of test function

Test function	f1	f2	f3	f4	f5
Dimension	2	2	2	5	5
Optimal distribution	Single peak	Multi peak	Complex multi peak	Multi peak	Complex multi peak
Variable range	[-5.12,5.12]	[-5.12,5.12]	[-500,500]	[-10,10]	[-600,600]
Global optimal solution	[1,1]	[0,0]	[420.9687, 420.9687]	[0,0,0,0,0]	[0,0,0,0,0]
Global optimal value	0	0	0	0	0
Accuracy requirements	1×10^{-3}	1×10^{-5}	3×10^{-1}	1×10^{-5}	1×10^{-3}

Table 2. Statistical table of reliability test results

Function	Method	Meet the accuracy requirements	Control rates	Best	Mean
f1	PSO	195	97.5%	1.020962173183369e-4	0.0136434802252842
	APSO	189	94.5%	1.22455851741843e-5	0.001001051395270
	BFPSO	198	99.0%	1.86337017700319e-13	0.000253597371737501
f2	PSO	181	90.5%	0	1.54116880610374 e-5
	APSO	162	81.0%	0	1.46635488190231 e-5
	BFPSO	199	99.5%	0	5.24142685911032 e-6
f3	PSO	178	89.0%	0.172065452841252	0.622747078460316
	APSO	185	92.5%	0.0136952403170199	0.177770974202361
	BFPSO	191	95.5%	2.54551323450869e-5	0.0206438668787996
f4	PSO	154	77.0%	8.16946510440175e-9	8.19113710888889 e-5
	APSO	160	80.0%	7.29585281078471e-9	3.88195236493864 e-5
	BFPSO	162	81.0%	9.36495325731812e-10	2.64480185942740 e-6
f5	PSO	162	81.0%	6.51823898750270 e-4	0.0654890233968797
	APSO	195	97.5%	2.03284256395264 e-8	0.00435214963708608
	BFPSO	195	97.5%	6.34048369363427e-16	0.00303619265847559

Table 3. Parameters comparison of different design methods

Design Method	m_{n1}	m_{n2}	z_1	z_3	i_1	β	Center Distance
Experience Design	3	5	19	17	6.3	11	470
Paper[7]	2	4	19	16	5.8	9.8	340
Present	2	4	17	16	5.9	10.1	324

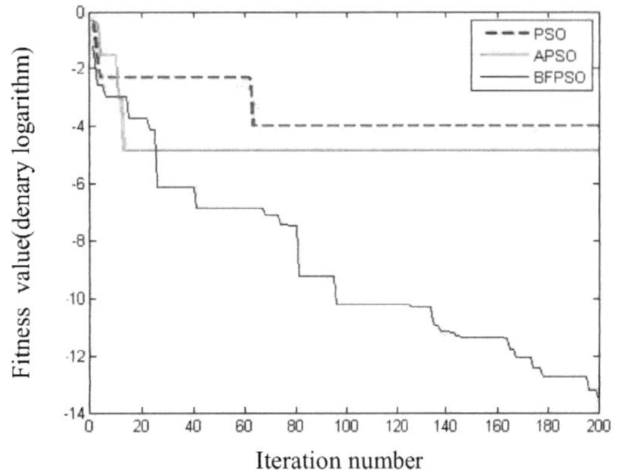

a. Convergence process of f1 function

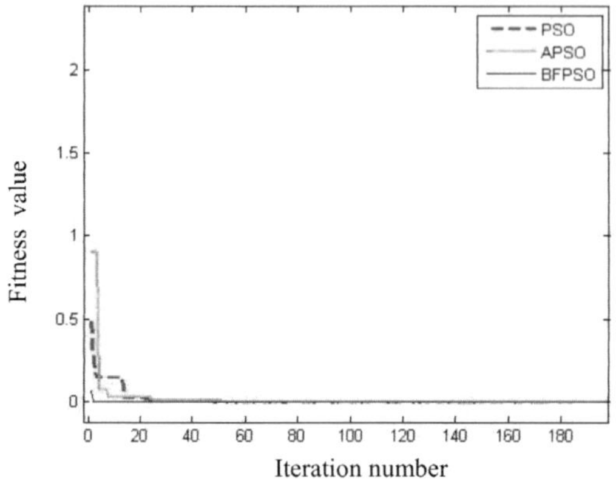

b. Convergence process of f2 function

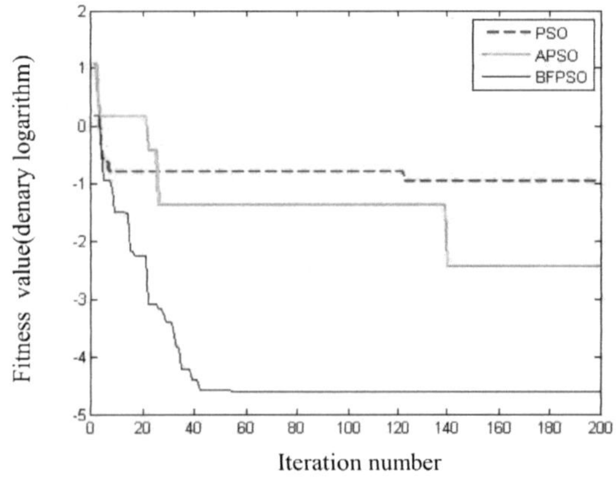

c. Convergence process of f3 function

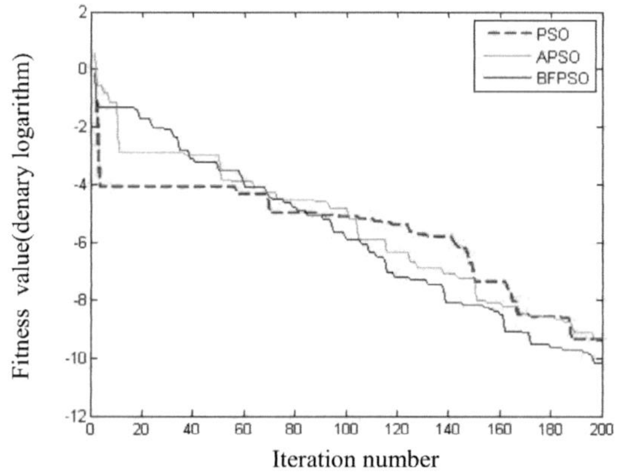

d. Convergence process of f4 function

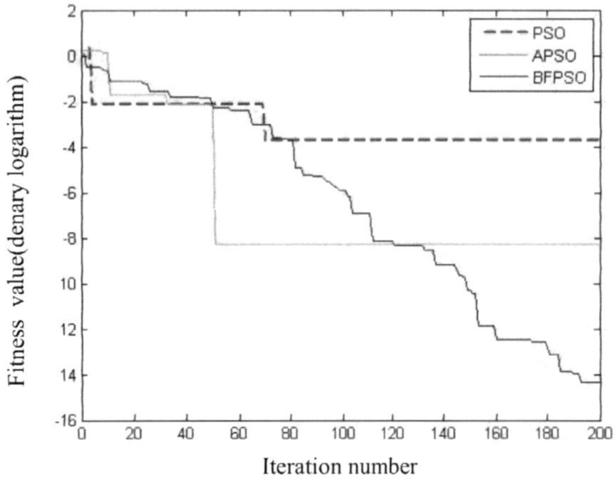

e. Convergence process of f5 function

Figure 2. Comparison of convergence results of the three algorithms

experience method and traditional optimization methods introduced in reference [7]. The comparative results are shown in Table 3, The convergence process of center distance varying with the number of iterations shown in Figure 3.

The conclusion can be drawn in Table 3 :the optimization result about center distance using BAPSO in this paper decreases 31.06% compared to experienced design,and 4.71% compared to traditional optimization methods, which reflects its capability of solving complex optimization problems. This means that using the improved optimization algorithm in this paper for mechanical design can improve the design accuracy and design efficiency. At the same time, it shows that this optimization algorithm can effectively deal with the practical problems in the design after it is introduced into the design field.

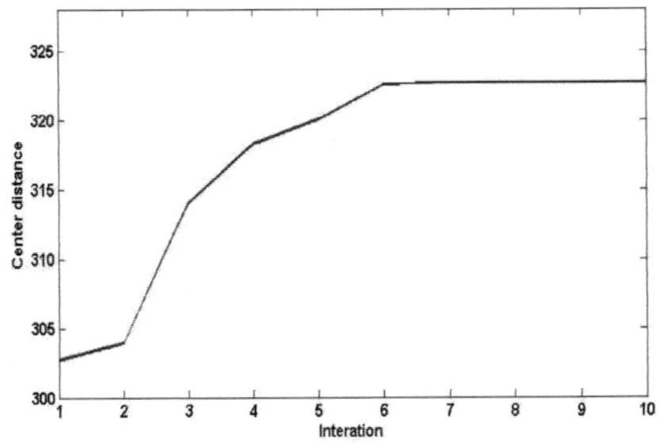

Figure 3. Iterative convergence process of center distance

5 Conclusions

In this paper, a particle swarm optimization algorithm based on bacterial foraging behavior(BFPSO) is proposed.

The performance of the BAPSO is tested by the classical test function, the results are shown that the BAPSO has better optimization ability, higher precision and reliability. Application the new algorithm for optimization design of gear reducer can get lesser center distance, the reducer structure is more compact. Indicating BAPSO has been more practical and better capacity for solving complex optimization problems. Further research in this area means that design accuracy and efficiency can be improved in solving practical problems in the areas of mechanical design, redundant allocation, flight control.

References

1. Deng X H. The method of optimizing design of cylindrical gear reducer[J].Journal of Minjiang University,2004，25（5）：38～40
2. Hu X H. Optimum design of single stage cylindrical gear reducer[J]. Combination Machine Tool and Automatic Machining Technology,2006(7)
3. Yang S Y, Zhang H. Group intelligence and bionic calculation[M]. Beijing. Electronic Industry Press, 2012
4. Yu S W. Case analysis and application of MATLAB optimization algorithm[M]. Beijing. Tsinghua University Press, 2014.
5. Hu J. Improvement of bacterial foraging optimization algorithm and its application[D]. Wuhan .Wuhan University of Technology, 2012.
6. Gong C,Wang Z L. Proficient in MATLAB Optimization[M]. Beijing. Electronic Industry Press, 2012
7. Zhang H P.optimization design of two-stage helical cylindrical gear reduce based on MATLAB[J]. Machinery Design and Manufacture,2010,(4):79-80.
8. Wen B C.Reducer and transmission[M].Beijing. China Machine Press,2015.
9. Wang D L,Ma Y L.Mechanical design[M]. Beijing.China Machine Press,2015.

AU DIY-Unicycle Balancing and Riding System (AU DIY-UBRS)

Narong Aphiratsakun , Thanakit Limpakhuptathaworn and Jiradech Kongthon

Vincent Mary School of Engineering, Assumption University, Thailand

Abstract. This paper presents and discusses the Assumption University unicycle (hereafter referred to as AU unicycle). The AU unicycle was built on campus, and it can work in two modes: balancing control and riding control. A gyroscope and an accelerometer sensor are used as sensors to sense the balancing position of the unicycle. A DC motor equipped with a gear set, sprockets, and a chain acts as an actuator to operate the wheel of the unicycle system. The main contribution and the objective of this paper are to demonstrate the implementation methodology and the balancing and riding control methods of the AU unicycle. The experiments on the AU unicycle show good results for both the balancing control method and the riding control method.

1 Introduction

The AU DIY-Unicycle Balancing and Riding System (AU DIY-UBRS) was designed and made based on a one-wheeled motorcycle, which must be able to balance itself for safe riding. It is important that a unicycle provides a rider with a safe ride. It was reported by Wang [1] that in the United States, an estimate of 3360 people were injured from unicycle riding from 1991 to 2010.

The RYNO (Ride Your New Opportunity) was invented by Tony Ozrelic in 2008, and it was equipped with a 6 axis gyroscope and an accelerometer working as sensors to balance the system. The RYNO has a leaning angle of approximately 15 degrees forward and backward, and it can travel with a speed of about 10 mph.

In September 2010, Solowheel, invented by Shane Chen, was presented in Las Vegas Bike Expo. Solowheel employed a gyroscope and an accelerometer as sensors to balance itself. Solowheel can move with a speed of around 12 mph.

Using a unicycle would be a good choice where parking space is limited [2].

Several research works on balancing systems have been done in the Vincent Mary School of Engineering at Assumption University, such as the Autonomous AU Bicycle: Self-Balancing and Tracking Control (AUSB2), [3]. The AUSB2 has used a gyroscope and an encoder as sensors for the system. Another work done by Aphiratsakun et al., [4] is on the implementation of the AU Balancing BallBot (AuB3). A gyroscope has been used as a sensor for the AuB3.

This paper presents the implementation methodology and the balancing and riding control methods of the AU DIY-UBRS made by engineers of the Vincent Mary School of Engineering at Assumption University. The system is in the form of a single-wheeled motorcycle. The AU DIY-UBRS has two modes: balancing mode and riding mode. The balancing mode is without a rider on the system while the riding mode is with a human as a rider.

The paper is structured as follows. Section 2 discusses the AU DIY-UBRS design including the design of mechanical parts and electronic components. The details of the control methods are explained in Section 3. The experiments are carried out in Section 4. Section 5 provides the results and discussions. The conclusions are discussed in Section 6..

2 AU DIY-UNICYCLE balancing and riding system

This section presents the AU-DIY-UBRS design for both mechanical parts and electronics components.

2.1 Mechanical design

The AU DIY-UBRS was designed by using SolidWorks. The design was based on a one wheel motorcycle. A 24V DC gear motor is used as a main actuator. The CAD design is shown in Fig. 1.

The original prototype AU DIY-UBRS is shown in Fig. 2. All of the equipment items can be obtained from the local market.

(a) (b)

Figure 1. AU DIY-UBRS CAD design
(a) Side View, (b) Isometric View

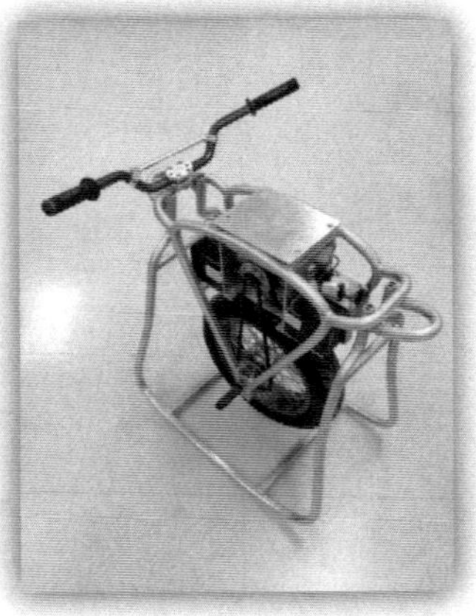

Figure 2. AU DIY-UBRS prototype

2.2 Electronics design

In this section, the controller, the sensor, and the actuator used in the AU unicycle are discussed.

2.2.1 Controller

An Arduino UNO R3 was chosen to be used as a controller for the AU DIY-UBRS. The Arduino UNO R3 is based on the ATmega328P 16 MHz quartz crystal, with 14 digital input/output pins, 6 analog inputs. The Arduino UNO R3 is shown in Fig. 3.

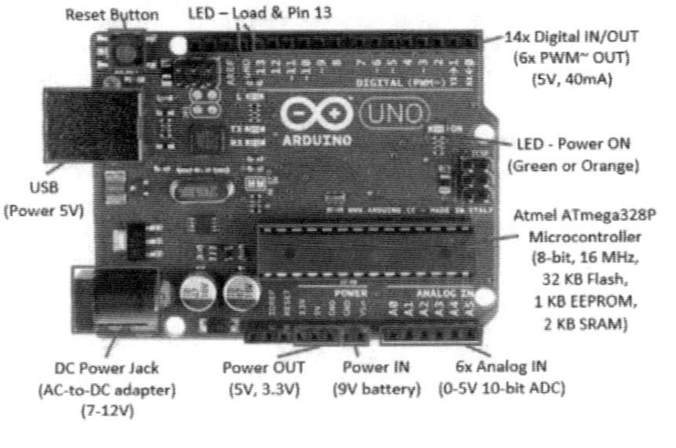

Figure 3. Arduino UNO R3

2.2.2 Sensor

Fig. 4 shows an MPU6050 gyroscope and an accelerometer sensor, which are used to sense the position of the system. The MPU6050 is composed of a 3-axis gyroscope and a 3-axis accelerometer on a digital motion processor that is capable of processing complex 6-axis motion fusion algorithms and is known as an inertial measurement unit.

The Arduino UNO R3 controller communicates with the MPU6050 through I2C function. Table 1 shows the specifications of the MPU6050.

Figure 4. MPU6050

Table 1. MPU Specifications

Specifications	Range
Accelerometer	±2, ±4, ±8, ±16g
Gyroscope	± 250, 500, 1000, 2000 °/s
Operation Voltage	3.3V-5.0V

2.2.3 Actuator

A 24V 350W DC Motor (MY1016YZ3) is used as an actuator for the AU DIY-UBRS. This motor has a speed and a rated torque of approximately 330 RPM and 9.47 Nm respectively. The motor is attached to a transmission gear of 9.8:1 ratio. The motor and the transmission gear are shown in Fig. 5. The motor-gear set is furnished with a motor driver (SE-HB-100). This motor driver works on the PWM concept. Table 2 shows the specifications, and Fig. 6 shows the motor driver (SE-HB-100).

Figure 5. DC motor and transmission gear

 https://doi.org/10.1051/matecconf/201816602005

Table 2. Motor driver (SE-HB-100) specifications

Specifications	**Range/Data**
Output	DC Motor 12-48V 80A (pulse current)
Input	Opto-isolated 5V 8mA TTL level
Drive mode	- On/Off control - Directional control - PWM control
PWM frequency	400-1000 Hz

Figure 6. Motor driver (SE-HB-100)

3 Control methods

Different control approaches have been taken to improve the performance, stability, and safety of unicycles. Many works on control developments of unicycles have been done. For example, Zenkov et al., [5] presented an approach to stabilizing the unicycle with a rider by imposing a feedback control force on the rider's limb. Buccieri et al., [6] proposed and implemented a control technique that uses a higher-dimensional state extension that can reject a disturbance on the unicycle. Franke et al., [7] used the motion planning and feedback control of a unicycle model. Jin et al., [8] proposed and verified a dynamics-based posture-balancing controller for a one-wheel pendulum robot. A fuzzy logic controller for a unicycle was designed by Xu et al., [9]. Do et al., [10] designed path tracking controllers for controlling unicycle-type mobile robots. Han et al., [11] employed a dynamic-model-based control method for the balancing and velocity control of a unicycle robot. An integral linear quadratic regulator and sliding mode controller on a unicycle robot was implemented by Moham et al., [12].

In this work, PID controllers are used to balance and stabilize the AU DIY-UBRS for both the balancing mode and the riding mode.

3.1. Balancing mode

A classical PID controller is implemented as a control method to balance the AU DIY-UBRS. In this mode, no rider is on the unicycle. The block diagram for the balancing control is illustrated in Fig. 7. The PID controller equations, as in Nise (2015), are shown in Equations (1), (2), (3), and (4) where Kp, Ki and Kd are a proportional gain, an integral gain and a derivative gain respectively. ERR is an error between a desired set point which is the balancing position and an actual tilt angle from the system.

$$PWM = (PID) \times (ERR) \tag{1}$$

$$PID = K_p + \frac{K_i}{s} + K_d s \tag{2}$$

$$EER = SETPOINT - ANGLE \tag{3}$$

$$DIR = \begin{matrix} 1 \;\; \text{if } ANGLE \text{ is negative or} \\ -1 \;\; \text{if } ANGLE \text{ is positive} \end{matrix} \tag{4}$$

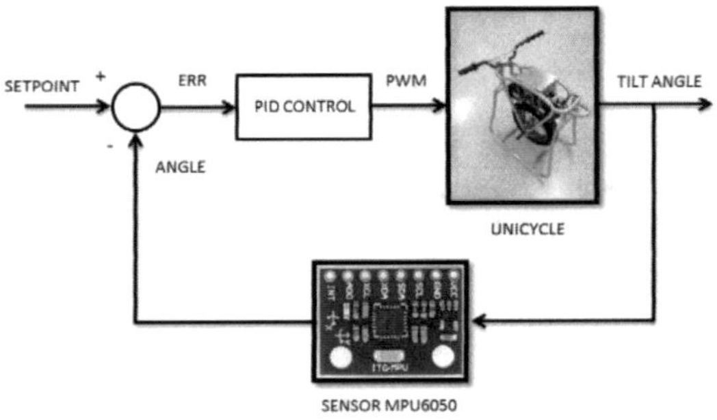

Figure 7. AU DIY-UBRS balancing block diagram

3.2 Riding mode

For the riding mode, a classical PID controller is also used as a controller for a person to ride on the AU DIY-UBRS. In this mode, a rider is on the system. The block diagram for the riding control is presented in Fig. 8. The human load acts as another input to the plant. This input is considered as a disturbance to the system. Equations (1), (2), (3), and (4) are still applied to the system. There is a reverse riding direction for this mode. Equation (5) shows that the AU DIY-UBRS will either go forward or stop for this mode. If a rider leans forward, the AU DIY-UBRS will go forward with a speed applied to the PWM function. If a rider leans backward, the AU DIY-UBRS will stop. This is also applied to the PWM function. The total PWM applied to the system is given in equation (6).

$$DIR = \begin{matrix} 1 \;\; \text{if } ANGLE \text{ is negative or} \\ 0 \;\; \text{if } ANGLE \text{ is positive} \end{matrix} \tag{5}$$

$$TOTAL\_PWM = PWM + DISTURBANCE \tag{6}$$

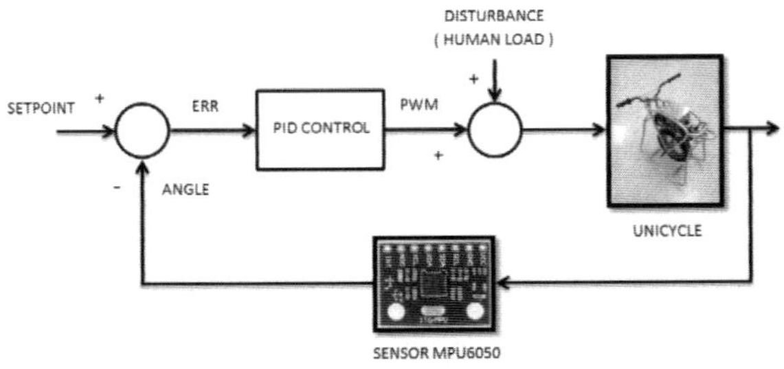

Figure 8. AU DIY-UBRS riding block diagram

4 Experiments

This section presents the experiments of the AU unicycle for both the balancing mode and the riding mode.

4.1 Balancing mode

In the experiment of the balancing mode, the experiment is carried out with PID gains tuned to Kp =10.5, Ki =0.8, Kd =8.55. There is no rider for the experiment. The setpoint is set to a horizontal balancing position.

4.2 Riding mode

In the experiment of the riding mode, the experiment is carried out with PID gains tuned to Kp =18.0, Ki =0.5, Kd =0.01. There is a rider for the experiment. The setpoint is set to a horizontal balancing position.

5 Results and discussions

This section reports the experiment results and provides the discussions on the experiment results.

5.1 Balancing mode

From the experiments, the plots of the angle (degree), error (degree), PWM (%), and direction versus time (second) are shown in Fig. 9 for the balancing mode.

It is shown in Fig. 9 that the response is very good, and the AU DIY-UBRS can balance itself for a very long period of time without falling.

5.2 Riding mode

From the experiments, the plots of the angle (degree), error (degree), PWM (%), and direction versus time (second) are shown in Fig. 10 for the riding mode.

In the riding mode, a rider acts as a disturbance to the system. It is shown in Fig. 10 that the response is very good and the AU DIY-UBRS can be ridden and moved forward if a rider leans forward, and it stops if a rider leans backward.

6 Conclusions

The AU DIY-Unicycle Balancing and Riding System (AU DIY-UBRS) was built, implemented, and tested successfully.

In the experiments, the balancing mode and the riding mode were carried out. In the balancing mode, PID gains were tuned to Kp =10.5, Ki =0.8, Kd =8.55. The response was very good, and the AU DIY-UBRS can balance itself for a very long period of time without falling. In the riding mode, a rider acts as a disturbance to the system. In the riding mode, PID gains were tuned to Kp =18.0, Ki =0.5, Kd =0.01. The AU DIY-UBRS can go forward if a rider leans forward, and it stops if a rider leans backward.

The AU DIY-UBRS was tested several times, and it could perform well for both modes. For the future work, an incremental encoder should be added to the system in order to sense the direction and the velocity of the system.

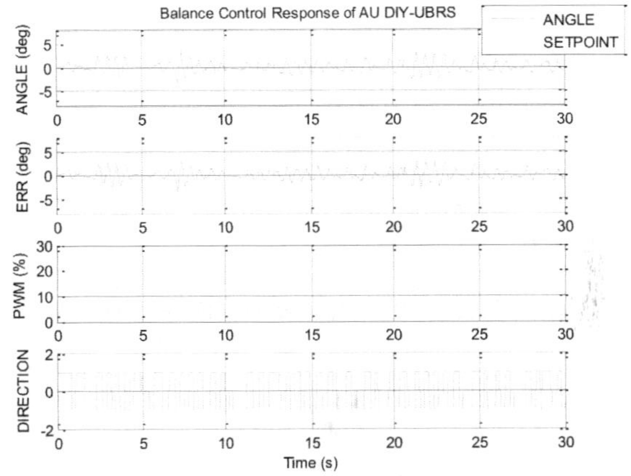

Figure 9. AU DIY-UBRS balancing control response

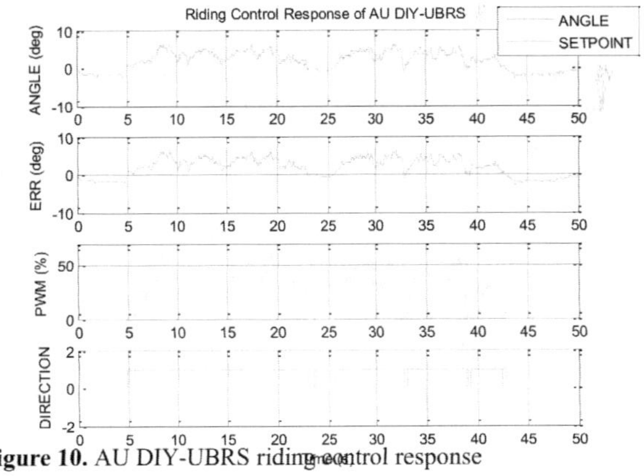

Figure 10. AU DIY-UBRS riding control response

References

1. Wang, M.L., (2013). Unicycle Injuries in the United States, The Journal of Emergency Medicine, Vol. 45, Issue 4, October 2013, pp 502–507.
2. Kleiner, K., (2011). Unicycle-Bot that can Balance while not Moving, NewScientist, Vol. 210, Issue 2816, 11 June 2011, pp. 22.
3. Aphiratsakun, N., Techakittiroj, K. (2013). Autonomous AU Bicycle: Self-Balancing and Tracking Control (AUSB2), Proceedings of the IEEE

International Conference on Robotics and Biomimetics (Robio 2013),pp. 480-485, 12-14 December 2013, Shenzhen, China.

4. Aphiratsakun, N., Nordeng, P. K. R., Suikkanen, M.,Lorpatanakasem, N. (2014). Implementation of AU Balancing Ball Bot (AUB3), Proceedings of the 2014 International Electrical Engineering Congress (iEECON2014), 17-19 March 2014, Pattaya, Thailand.
5. Zenkov, D.V., Bloch, A.M., Marsden, J.E.,(2002). Stabilization of the Unicycle with Rider, Proceedings of the 38th IEEE Conference on Decision and Control, Cat. No.99CH36304, Date Added to IEEE Xplore: 06 August 2002, Date of Conference: 7-10 December 1999.
6. Buccieri, D., Perritaz , D., Mullhaupt, P., Jiang, Z.-P., Bonvin, D.,(2009). Velocity-Scheduling Control for a Unicycle Mobile Robot: Theory and Experiments, IEEE Transactions on Robotics, Vol. 25, Issue 2, April 2009, pp. 451 – 458.
7. Franke, M., Rudolph, J., Woittennek, F., (2009). Motion Planning and Feedback Control of a Planar Robotic Unicycle Model, IFAC Proceedings Volumes, Vol. 42, Issue 13, 2009, pp. 501-506.
8. Jin, H., Hwang, J., Lee, J., (2011). A Balancing Control Strategy for a One-Wheel Pendulum Robot Based on Dynamic Model Decomposition: Simulations and Experiments, IEEE/ASME Transactions on Mechatronics, Vol.16, Issue 4, August. 2011, pp. 763 – 768.
9. Xu, J., X., Guo, Z., Q., Lee, T. H., (2012). Synthesized Design of a Fuzzy Logic Controller for an Underactuated Unicycle, Fuzzy Sets and Systems, Vol. 207, 16 November 2012, pp. 77-93.
10. Do, K.D., (2013). Bounded Controllers for Global Path Tracking Control of Unicycle-Type Mobile Robots, Robotics and Autonomous Systems, Vol. 61, Issue 8, August 2013, pp. 775–784.
11. Han, S. I., Lee, J. M. , (2015). Balancing and Velocity Control of a Unicycle Robot Based on the Dynamic Model, IEEE Transactions on Industrial Electronics ,Vol. 62, Issue 1, January. 2015,pp.405 – 413.
12. Moham, S., Nandagopal, J.L., Amritha ,S.,(2016). Decoupled Dynamic Control of Unicycle Robot Using Integral Linear Quadratic Regulator and Sliding Mode Controller, Procedia Technology, Vol.25, 2016, pp. 84-91.

Modular Estimation Strategy of Vehicle Dynamic Parameters for Motion Control Applications

Mustafa Rawash , Mohamed Abdelaziz[1], Maged Ghoneima[2] and Farid Tolbah[3]

[1]*Ain Shams University, Faculty of Engineering, Automotive Department, Cairo, Egypt*
[2]*Ain Shams University, Faculty of Engineering, Mechatronics Department, Cairo, Egypt*
[3]*Ain Shams University, Faculty of Engineering, Design and Production Department, Cairo, Egypt*

Abstract. The presence of motion control or active safety systems in vehicles have become increasingly important for improving vehicle performance and handling and negotiating dangerous driving situations. The performance of such systems would be improved if combined with knowledge of vehicle dynamic parameters. Since some of these parameters are difficult to measure, due to technical or economic reasons, estimation of those parameters might be the only practical alternative. In this paper, an estimation strategy of important vehicle dynamic parameters, pertaining to motion control applications, is presented. The estimation strategy is of a modular structure such that each module is concerned with estimating a single vehicle parameter. Parameters estimated include: longitudinal, lateral, and vertical tire forces – longitudinal velocity – vehicle mass. The advantage of this strategy is its independence of tire parameters or wear, road surface condition, and vehicle mass variation. Also, because of its modular structure, each module could be later updated or exchanged for a more effective one. Results from simulations on a 14-DOF vehicle model are provided here to validate the strategy and show its robustness and accuracy.

1 Introduction

The presence of active safety systems, such as Antilock Braking, Traction Control and Stability Control systems, have become increasingly important. Such systems or motion control systems help maintain driver's control over the vehicle in dangerous driving situations, thereby reducing accident rates. Also such systems are capable of enhancing vehicle performance and handling.

The effectiveness of such control systems can be enhanced, if combined with knowledge of certain dynamic parameter. Some of these parameters are difficult to measure due to either technical or economic reasons, therefore, the only practical alternative is estimation of these parameters.

Typically, vertical tire forces are estimated using longitudinal and lateral load transfers, and static loads on each wheel [1-4]. In [5] Antonov et al. proposed an advanced method for calculating vertical tire forces based on modelled suspension system and using estimated forces acting on vehicle centre of gravity.

Longitudinal or lateral tire forces can be estimated based on analytical tire models, such as linear model [6], Dugoff model [7], or Magic Formula model [8]. However, such models rely on parameters that are specific to each tire and vary with tire aging or road surface condition. Alternatively, estimation of longitudinal tire forces based on modelling wheel rotational dynamics and vehicle longitudinal dynamics has been pursued in the literature [1,9]. In [1], Cho et al. defined a Lyapunov function based on wheel rotational dynamics, to estimate longitudinal tire force by satisfying the Lyapunov stability conditions. A random-walk model was also defined for combined (longitudinal and lateral) tire force estimation. In [10] and [11] a sliding mode observer is used.

Lateral tire forces are more difficult to estimate than longitudinal tire forces. A number of studies have tried to estimate the summation of lateral tire forces on front wheels and rear wheels using yaw-plane bicycle models [1, 2, 12, 13, 14]. The summation can be further distributed on left and right wheels using normal forces acting on each wheel.

Knowledge of vehicle mass is important to vehicle dynamic parameters estimation. Moreover, vehicle mass could change significantly with the number of passengers in it or the amount of luggage. In [3,15], vehicle mass identification is made at rest using suspension deflection transducers, which are uncommon in vehicles without active or semi-active suspension. Many authors utilize the fact that vehicle mass directly affects the relationship between the net longitudinal force on the vehicle and its longitudinal acceleration to design vehicle mass estimators. In [16], Bae et al. proposed an averaging recursive least squares estimator that uses longitudinal force, acceleration, and GPS-based road grade measurements to determine vehicle mass and aerodynamic drag. In [17], Grieser proposed a similar averaging estimator in which aerodynamic drag is simulated online, rather than estimated, then subtracted from force measurements. In [18, 19], Vahidi et al. proposed a similar averaging estimator that does not require road grade measurements, rather estimates vehicle mass, drag, and road grade simultaneously. The system accommodates the time-varying nature of aerodynamic drag and road grade through multi-rate forgetting. Also in [20], Winstead and Kolmanovsky used an extended Kalman filter for estimating vehicle states including mass.

There exist several tire force estimation systems in the literature [1-5, 11-13, 21, 22]. However, most of them suffer from dependence on knowledge of vehicle mass [1,4,5,10-13,21,22], dependence on suspension deflection transducers for estimating vehicle mass and body roll angle [2,3], or dependence on tire models which has the disadvantage mentioned earlier. Also most tire forces estimation systems found in the literature do not provide a unified structure for estimating tire forces in all directions. However, in [14], Rezaeian et al. presented a novel tire force estimation strategy. The strategy is of modular structure, where each module is concerned with estimating tire force in a specific direction, and each module could be further developed. Also the strategy has the advantage of independence of tire parameters or wear, as well as road surface condition and vehicle mass variation. Other parameters estimated in that strategy were vehicle mass and longitudinal velocity – body roll and pitch angles.

Motivated by the work in [14], this paper presents further enhancements to the aforementioned strategy as follows:

• Applying wheel slip ratio estimation for better evaluation of longitudinal velocity when wheels are slipping.
• Adopting a new vehicle mass estimation system; one that is more or equally effective, while requiring minimal instrumentation and computational cost.

For brevity of this paper, only developed modules will be discussed in details. For more information on the rest of the modules, please refer to [14]. The rest of this paper is structured as follows: Section-2 presents the outline of the modular estimation strategy with details on developed modules. Section-3 presents and discusses results of simulations carried to validate the proposed strategy. Finally, a conclusion is presented in Section-4.

2 Proposed estimation strategy

Figure 1 shows the outline of the proposed estimation strategy, and Table 1 lists all the symbols used in the figure (The subscript ij denotes each individual wheel). The modules of the estimation strategy are organized in a hierarchy, where each module relies on sensory signals and estimates from higher order modules.

Undeveloped modules are implemented as in [14] as follows:

• Each individual longitudinal tire force is estimated using an observer based on rotational wheel dynamics, and the force is estimated to satisfy a Lyapunov stability condition.
• Vertical tire forces on front and rear axles are first calculated using a 1-DOF pitch model and longitudinal and vertical acceleration measurements, then each front and rear axle forces are distributed between right and left tires using 1-DOF roll plane model and lateral and vertical acceleration measurements.
• Vertical tire forces estimation relies on estimates of body pitch and roll angles, which are estimated using observers similar to that initially proposed in [23] for body roll angle estimation, and later generalized in [14] for both body roll and pitch angles estimation. Body roll observer is based on 1-DOF roll plane model and relies on lateral acceleration roll rate measurements, while body pitch observer is based on 1-DOF pitch plane model and relies on longitudinal and pitch rate measurements.
• Lateral tire forces on front and rear axles are first estimated using an Unscented-Kalman-Filter as an observer for 3-DOF yaw plane bicycle model, where the forces are modelled as random walk states. Then each axle forces are distributed between right and left tires using vertical load distribution.

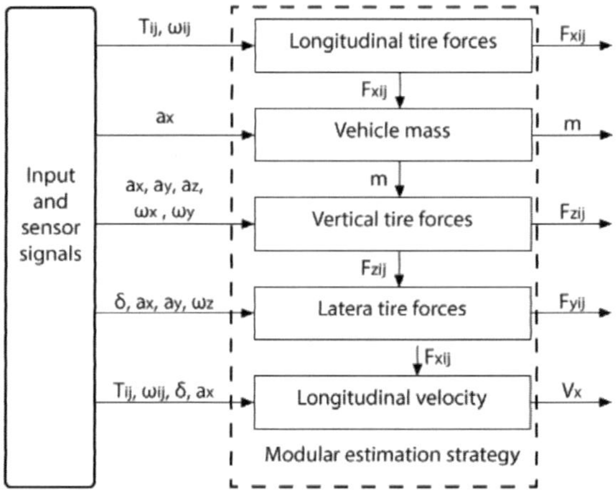

Figure 1. Estimation strategy outline.

Table 1. Definitions of symbols used in Figure 1.

Symbol	Definition
T_{ij}	Wheel driving or braking torque
ω_{ij}	Wheel angular velocity
δ	Front steering angle
a_x	Measured longitudinal acceleration
a_y	Measured lateral acceleration
a_z	Measured vertical acceleration
ω_x	Measured roll angular velocity
ω_y	Measured pitch angular velocity
ω_z	Measured yaw angular velocity
F_{xij}	Estimated longitudinal tire forces
F_{yij}	Estimated lateral tire forces
F_{zij}	Estimated vertical tire forces
m	Estimated vehicle mass
V_x	Estimated longitudinal velocity

2.1 Vehicle mass estimation

In [24], Fathy et al. proposed a mass estimation system that relies on minimal instrumentation, as opposed to previous endeavours by authors where mass estimation relies on knowing vehicle aerodynamic characteristics or road grade angle [16,17]. Additionally, the proposed system is simpler than the ones presented by Vahidi et al. in [18, 19], Winstead and Kolmanovsky in [20]. Also in [24] the system was compared to those proposed by Bae et al. in [16] and Vahidi et al. in [18,19] to show faster and more accurate, reliable, and robust estimation.

The same system is applied here to estimate vehicle mass. It relies on the fact that inertial dynamics dominate the longitudinal behaviour of a vehicle at increasing frequencies. So when the vehicle motion is dominantly longitudinal the high frequency component of that motion obeys:

$$m.\delta a_x = \delta F_x \tag{1}$$

Where m is the vehicle mass, δa_x is the high frequency component of vehicle longitudinal acceleration and δF_x is the high frequency component of the net longitudinal forces acting on the vehicle. Lead-lag filters are used to extract high frequency components of longitudinal acceleration measurement and total longitudinal tire forces estimate.

The first part of any vehicle journey usually includes enough longitudinal excitation, so mass estimation could be carried out only until change in estimated mass is less than a predefined value, as suggested in [14]. In [24], to insure that the vehicle motion is predominantly longitudinal, a supervisory algorithm was proposed, where mass estimation is only carried out when certain conditions are met.

2.2 Longitudinal velocity estimation

In [14] the linear velocity of each wheel centre is calculated from its angular velocity measurement, then transferred to vehicle centre of gravity using measured yaw rate. All 4 velocities are averaged to obtain an estimate of vehicle's longitudinal velocity. If the angular acceleration of a wheel is more than a predefined value, indicating that such wheel might be slipping, it is omitted from averaging.

Such technique of estimating longitudinal velocity doesn't provide accurate velocity estimate since angular acceleration is not a reliable indication of wheel slippage, each wheel is prone to a slight value of slippage which affects final velocity estimate, and slipping of all 4 wheels will result in the absence of velocity estimate.

In this paper, it is intended to combine such technique of longitudinal velocity estimation with wheel slip estimation to obtain a more accurate velocity estimate, where the linear velocity of each wheel centre shall be additionally calculated based on estimated slip ratio.

In [25] a system was proposed for estimating individual wheel slip ratio using Extended-Kalman-Filter (EKF) without detecting vehicle speed. The same technique shall be applied here but while using an Unscented-Kalman-Filter (UKF) instead.

The slip ratio can be defined as:

$$\lambda = \frac{\omega.r - v_x}{v_x} \tag{2}$$

Where ω is the wheel angular velocity, r is the wheel radius, assumed to be known, and v_x is the wheel linear velocity at its center. From (2) the rate of change of slip ratio can be written as:

$$\dot{\lambda} = \frac{\dot{\omega}}{\omega}.(\lambda + 1) - \frac{\dot{v}_x}{\omega.r}(\lambda + 1)^2 = f(\lambda) \tag{3}$$

Where $\dot{v}_x$ is the linear acceleration of the wheel centre, and is assumed in this paper to be equal to the measured longitudinal acceleration of the vehicle. However, a more accurate value may be obtained from longitudinal acceleration, lateral acceleration, and steering angle measurements.

The slip ratio shall be regarded as the state, and a discrete form of state update equation for the UKF can be derived using forward Euler discretization of (3) as follows:

$$x_{k+1} = x_k + \Delta t.f(x_k) + w \tag{4}$$

Where x_{k+1} is the state at the current time step, x_k is the state at the previous time step, Δt is the sampling time, and w is a process noise with zero mean and covariance Q.

The wheel rotational dynamics can be written as:

$$T = F_z.C(\lambda).r + I.\dot{\omega} \tag{5}$$

Where T is the total driving or braking torque applied to the wheel, F_z is the vertical tire force estimated, $C(\lambda)$ is the longitudinal friction coefficient as a function of slip ratio, and I is the wheel moment of inertia.

In [25], $C(\lambda)$ was obtained from a Pacejka-Magic Formula tire model, which is against the strategy adopted in this paper because tire models rely on parameters that are specific to each tire and vary with tire aging or road surface condition. However, a modal representation of the coefficient of friction as a function of slip was applied, as shown in Figure 2, and later simulated on a wet road to verify its effectiveness.

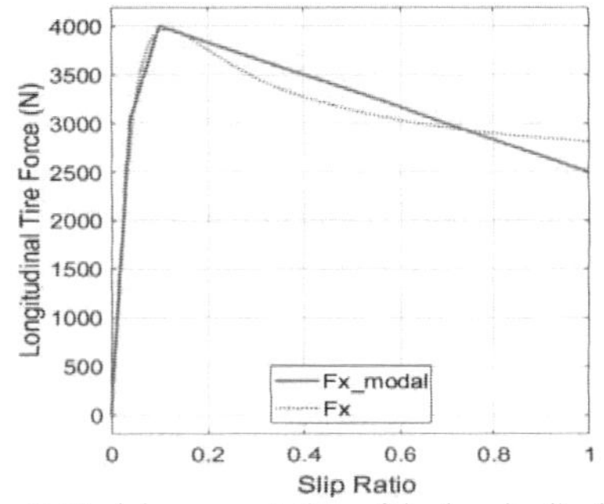

Figure 2. Modal representation of tire longitudinal force.

The observation equation for the UKF is obtained from (5), and the total applied torque is regarded as the measurement.

$$y_k = F_{z_k}.C(x_k).r + I.\dot{\omega}_k + v \quad (6)$$

Where v is a measurement noise with zero mean and covariance R. More information on UKF can be found in [26].

3 Simulation results

In [27], T. Shim and C. Ghike presented a 14-DOF vehicle model for vehicle dynamic simulations. The vehicle model was compared in 3 different simulated manoeuvres with 2 automotive industry standard Multi-body simulation software; namely CarSim and ADAMS/Car, and has shown very close output.

The same 14-DOF model is adopted here for validating the presented estimation strategy. The vehicle model was equipped with Magic formula tire model found in [8] using TNO MF-Tire/MF-Swift 6.1 Tire Model Parameters, and implemented in MATLAB along with the estimation strategy. The vehicle model was simulated as a 4-wheel-drive in-wheel-motor electric vehicle, where each wheel is equipped with an electric motor capable of applying instant torque. The steering system was simulated as a simple front Ackerman steering system with steering ratio of 1/15. For brevity of this paper more information on the vehicle or tire models and their parameters can be found in [27] and [8] respectively.

Sensory signals are needed for the estimation strategy, and include: longitudinal, lateral, and vertical accelerations (accelerometer), roll, pitch and yaw rate measurements (gyroscope), individual wheel rotational velocities, and front steering angle. All sensory signals were sampled from the vehicle model at a frequency of 100Hz, except for accelerations and body rotational velocities, which were additionally augmented with the effect of gravitational acceleration and sensor noise. The following equations describe the accelerometer and gyroscope sensor models.

$$\omega_{x_m} = \omega_x + \delta_1 \quad (7)$$

$$\omega_{y_m} = \omega_y + \delta_1 \quad (8)$$

$$\omega_{z_m} = \omega_z + \delta_1 \quad (9)$$

$$a_{x_m} = \dot{u} + \omega_y.w - \omega_z.v - g.sin\theta + \delta_2 \quad (10)$$

$$a_{y_m} = \dot{v} + \omega_z.u - \omega_x.w + g.cos\theta.sin\varphi + \delta_2 \quad (11)$$

$$a_{z_m} = \dot{w} + \omega_x.v - \omega_y.u - g.cos\theta.cos\varphi + \delta_2 \quad (12)$$

Where ω_{x_m}, ω_{y_m} and ω_{z_m} are the measured values of roll, pitch and yaw rates respectively with respect to body fixed frame, ω_x, ω_y and ω_z are the actual roll, pitch and yaw rates respectively with respect to body fixed frame, a_{x_m}, a_{y_m} and a_{z_m} are measured longitudinal, lateral and vertical accelerations respectively with respect to body fixed frame, u, v and w are longitudinal, lateral and vertical velocities respectively with respect to body fixed frame, g is the gravitational acceleration, θ and φ are pitch and roll angles respectively of the body fixed frame with respect to world frame, and finally δ_1 and δ_2 are zero mean random noises with variances 0.01 and 0.0003475 respectively.

As in [14], 3 manoeuvres where simulated for validating the estimation strategy; slalom manoeuvre on a dry road, slalom manoeuvre on a wet road, and a Double-Lane-Change (DLC) manoeuvre on dry road. The coefficient of friction for a dry road is 1, while for a wet road it's 0.6, also an additional 200 Kg were added to the actual vehicle mass in the slalom manoeuvre on wet road to evaluate the mass estimation module.

In all 3 manoeuvres, the vehicle is accelerated to 55km/h for the first 5 seconds, which is enough longitudinal excitation to obtain vehicle mass estimate. A switch was implemented to turn off mass estimation after the first 5 seconds have passed, and applying steering angles. Also in the last part of each manoeuvre the vehicle is decelerated.

The following subsections provide the simulation results of each manoeuvre, where the performance of each module is evaluated by calculating Normalized-Root-Mean-Squared (NRMS) error, which is calculated according to the following equation:

$$NRMS = \frac{\sqrt{\sum_{i=1}^{N}(A_{act} - A_{est})^2/N}}{\max(|A_{act}|)} \quad (13)$$

Where A_{act} is the actual signal, A_{est} is the estimated signal and N is the number of collected samples. Also samples were collected at a frequency of 100Hz.

3.1. Slalom manoeuvre on dry road

Figure 3 shows the input torque, front steering angle, measured longitudinal acceleration, and measured lateral acceleration of the slalom manoeuvre on a dry road.

Figures 3-7 show the outputs of all estimation modules compared with their real values. Figure 7 shows that the mass estimate converges very closely to the real value, while the percentage error falls below 1% in less than 2.5 seconds. It is to be noted that an upper limit was imposed on mass estimate, and the saturation limit was 700Kg above original vehicle mass, and also in Figure 6 the estimated vertical tire forces are inaccurately estimated during the first 2.5 seconds, which is the time taken for the mass estimate to converge to an accurate value.

Also Figure 8 shows the longitudinal velocity estimate of the new proposed module to be more accurate, while the former estimation module deviates significantly during slippage. Table 2 lists the calculated NRMS of all estimates and the percentage error calculated for the mass estimate.

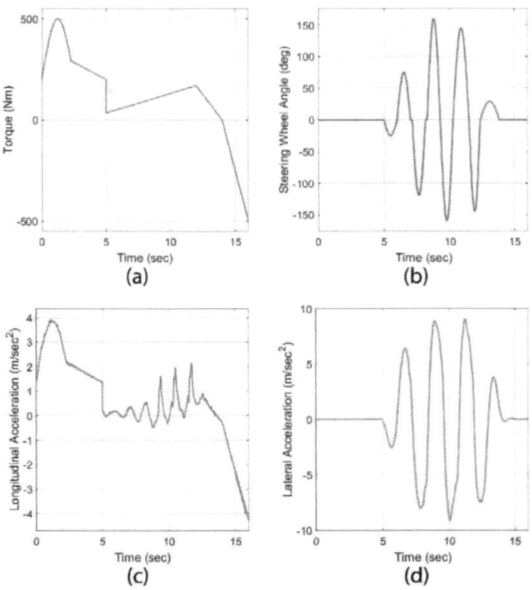

Figure 3. Slalom manoeuvre on dry road (a) input torque to each wheel (b) steering wheel angle (c) measured longitudinal acceleration (d) measured lateral acceleration.

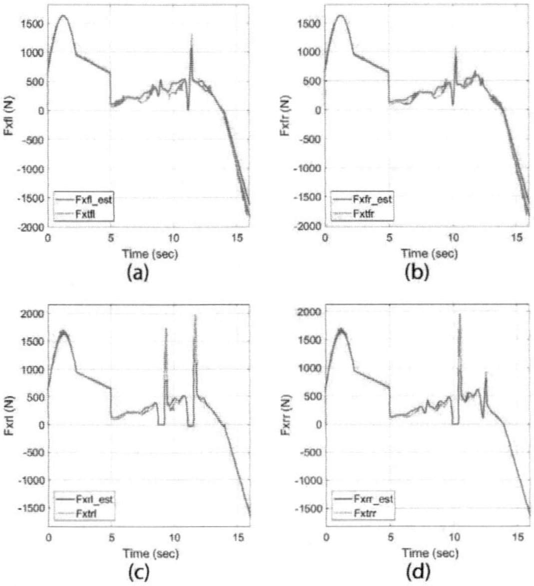

Figure 4. Estimated longitudinal forces in slalom manoeuvre on dry road (a) front-left (b) front-right (c) rear-left (d) rear-right. Fxtij denotes actual tire force, while Fxij_est denotes estimated tire force.

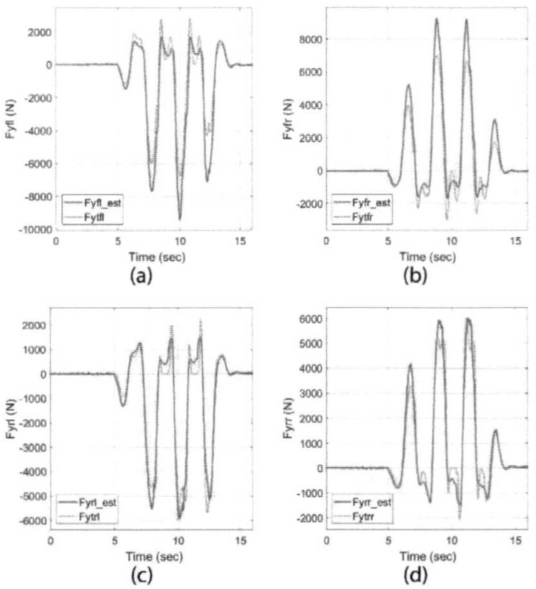

Figure 5. Estimated lateral forces in slalom manoeuvre on dry road (a) front-left (b) front-right (c) rear-left (d) rear-right. Fytij denotes actual tire force, while Fyij_est denotes estimated tire force.

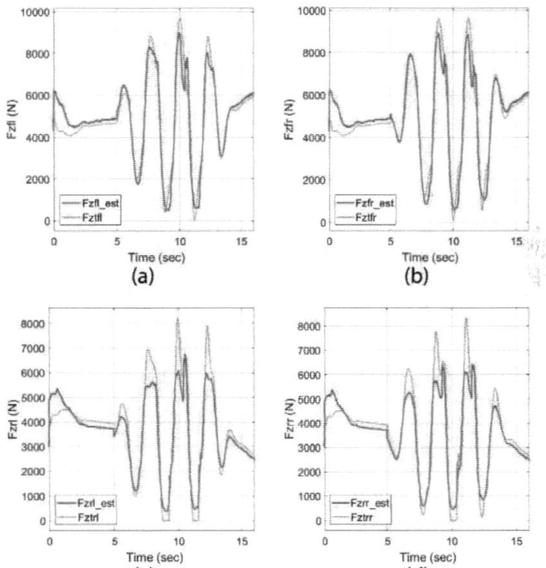

Figure 6. Estimated vertical forces in slalom manoeuvre on dry road (a) front-left (b) front-right (c) rear-left (d) rear-right. Fztij denotes actual tire force, while Fzij_est denotes estimated tire force.

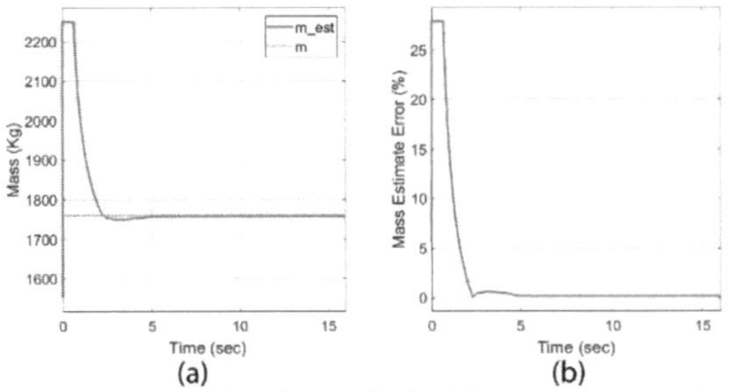

Figure 7. Mass estimation results in slalom manoeuvre on dry road. (a) comparison between actual mass m and estimated mass m_est. (b) percentage error for mass estimation.

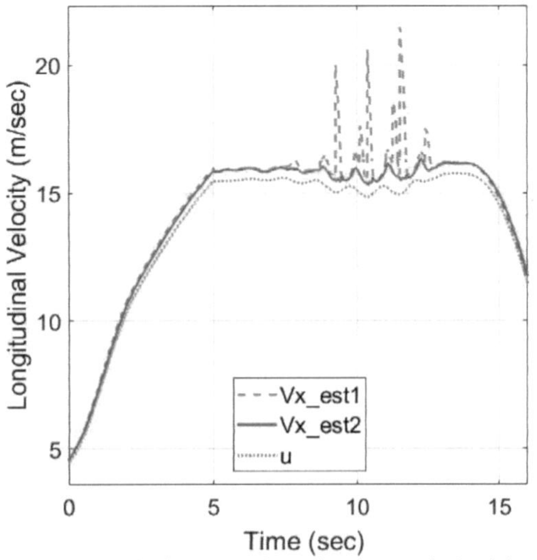

Figure 8. Longitudinal velocity estimation results in slalom manoeuvre on dry road. u denotes actual vehicle velocity, Vx_est1 denotes velocity estimation without slip estimation, Vx_est2 denotes velocity estimation with slip estimation.

Table 2. NRMS and maximum values for all modules outputs in slalom manoeuvre on dry road.

		FL	FR	RL	RR	N/A
F_x	NRMS	4.29	4.17	5.2	4.19	
	Max.	1791	1794	1985	1953	
F_y	NRMS	12.55	11.15	6.66	7.27	
	Max.	6727	6993	5979	6018	
F_z	NRMS	5.9	6.22	7.45	7.33	
	Max.	9693	9621	8218	8336	
V_x	NRMS					2.69
	Max.					15.77
m	PE					0.13

3.2 Slalom manoeuvre on wet road

Figure 9 shows the input torque, front steering angle, measured longitudinal acceleration, and measured lateral acceleration of the slalom manoeuvre on a wet road.

Figures 10-14 show the outputs of all estimation modules compared with their real values. Figure 13 shows that the mass estimate converges very closely to the real value which is 200Kg more than in manoeuvres on dry road, while the percentage error falls to about 2% in less than 2.5 seconds and eventually remains below 2%. Also Fig. 14 shows the new longitudinal velocity module still maintaining a more accurate output than the former estimation module. Table 3 lists all calculated NRMS values and percentage error for mass estimate.

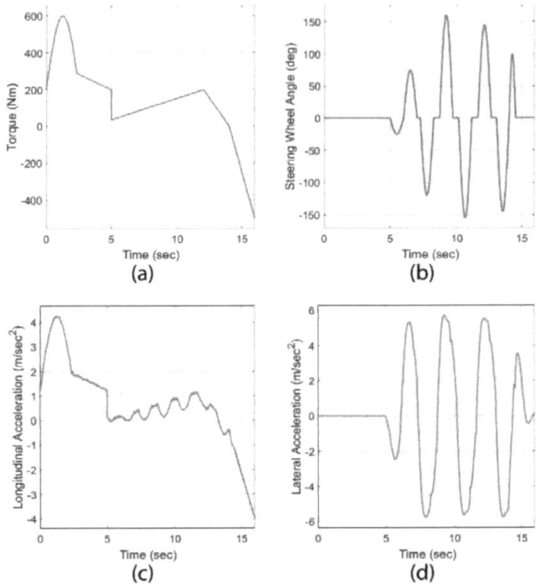

Figure 9. Slalom manoeuvre on wet road (a) input torque to each wheel (b) steering wheel angle (c) measured longitudinal acceleration (d) measured lateral acceleration.

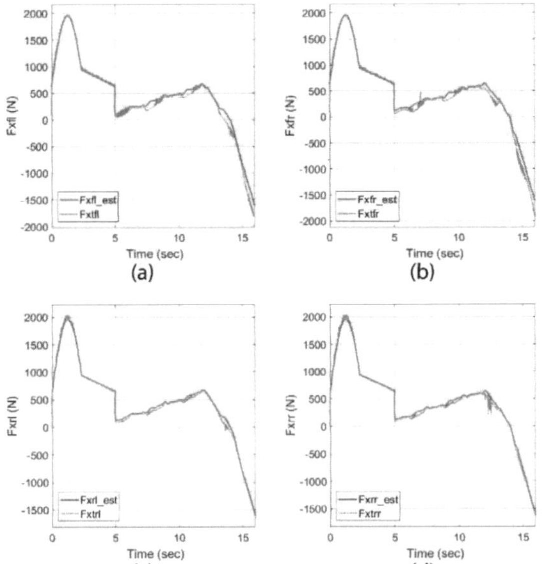

Figure 10. Estimated longitudinal forces in slalom manoeuvre on wet road (a) front-left (b) front-right (c) rear-left (d) rear-right. Fxtij denotes actual tire force, while Fxij_est denotes estimated.

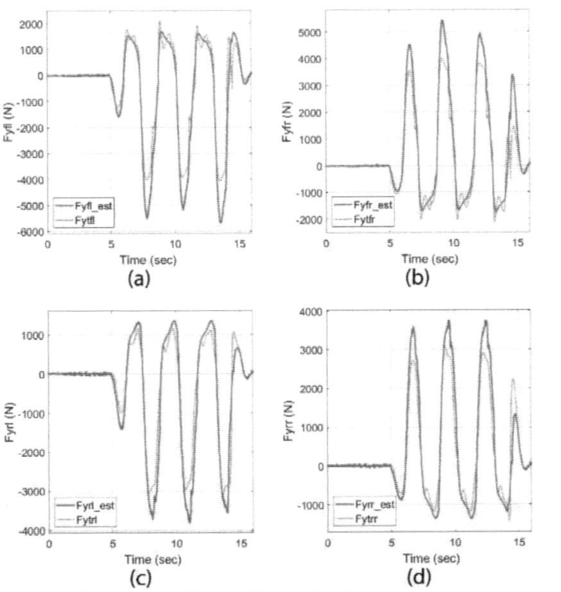

Figure 11. Estimated lateral forces in slalom manoeuvre on wet road (a) front-left (b) front-right (c) rear-left (d) rear-right. Fytij denotes actual tire force, while Fyij_est denotes estimated tire force.

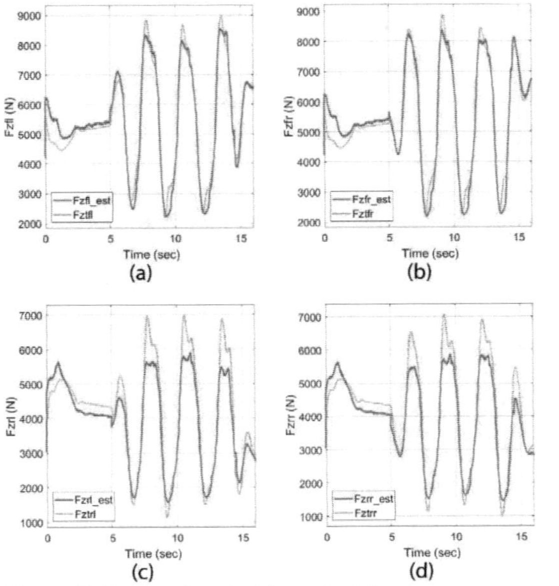

Figure 12. Estimated vertical forces in slalom manoeuvre on wet road (a) front-left (b) front-right (c) rear-left (d) rear-right. Fztij denotes actual tire force, while Fzij_est denotes estimated tire force.

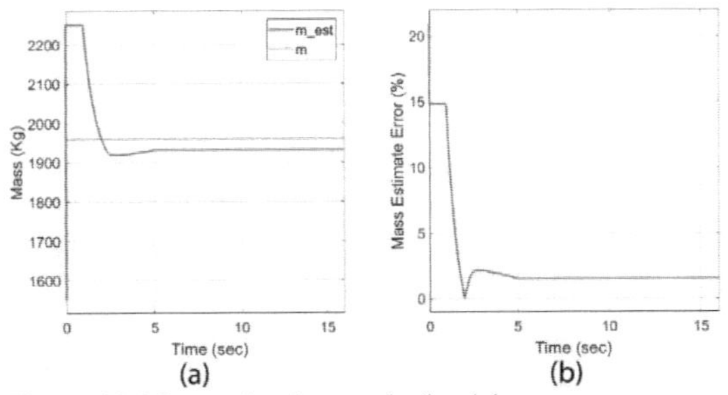

Figure 13. Mass estimation results for slalom manoeuvre on wet road. (a) comparison between actual mass m and estimated mass m_est. (b) percentage error for mass estimation.

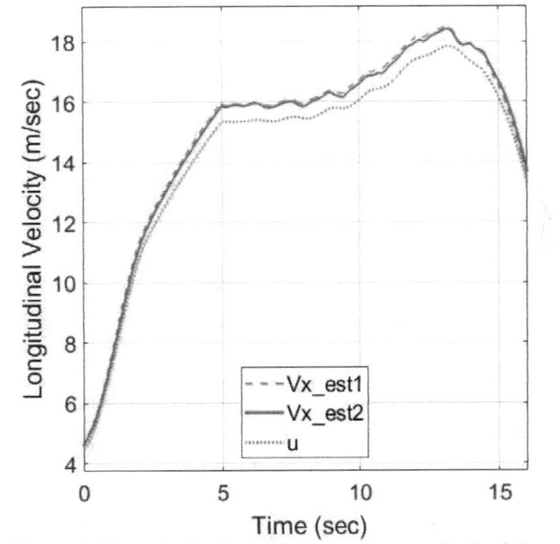

Figure 14. Longitudinal velocity estimation results in slalom manoeuvre on wet road. u denotes actual vehicle velocity, Vx_est1 denotes velocity estimation without slip estimation, Vx_est2 denotes velocity estimation with slip estimation.

Table 3. NRMS and maximum values for all modules outputs in slalom manoeuvre on wet road.

		FL	FR	RL	RR	N/A
F_x	NRMS	4.07	4.24	2.09	2.15	
	Max.	1973	1973	2034	2034	
F_y	NRMS	13.08	13.43	12.13	13.63	
	Max.	4079	4011	3083	3065	
F_z	NRMS	4.01	4.16	7.02	6.7	
	Max.	8989	8883	6999	7080	
V_x	NRMS					2.73
	Max.					17.83
m	PE					1.49

3.3 DLC manoeuvre on dry road

Figure 15 shows the input torque, front steering angle, measured longitudinal acceleration, and measured lateral acceleration of the DLC manoeuvre on a dry road.

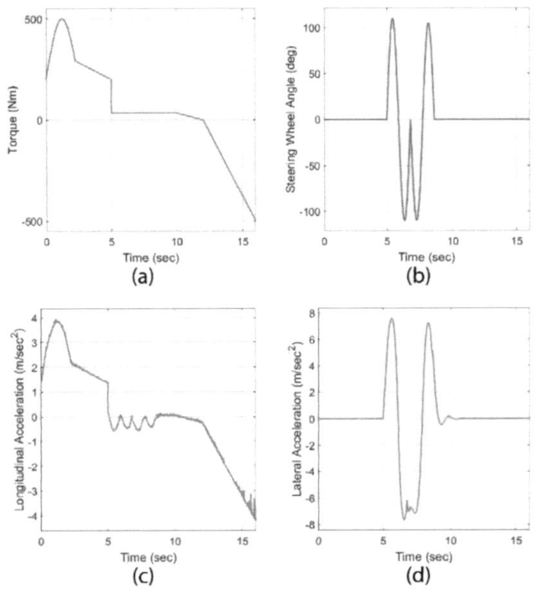

Figure 15. DLC manoeuvre (a) input torque to each wheel (b) steering wheel angle (c) measured longitudinal acceleration (d) measured lateral acceleration.

Figures 16-20 show the outputs of all estimation modules compared with their real values. Figure 19 shows the mass estimation module output to be exactly the same as in the slalom manoeuvre on dry road, while Figure 20 shows the new longitudinal velocity module to be still performing well. Table 4 lists all calculated NRMS values and percentage error for mass estimate.

The accuracy and performance of all undeveloped modules in all 3 manoeuvres is comparable to that found in [14].

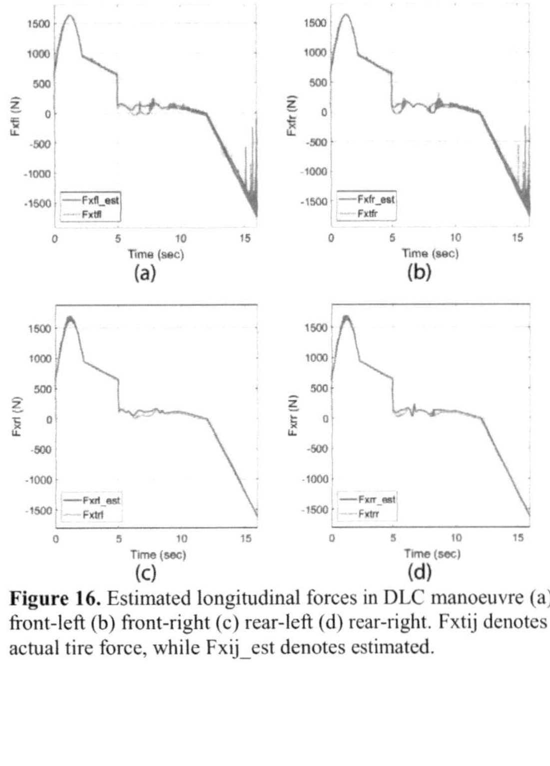

Figure 16. Estimated longitudinal forces in DLC manoeuvre (a) front-left (b) front-right (c) rear-left (d) rear-right. Fxtij denotes actual tire force, while Fxij_est denotes estimated.

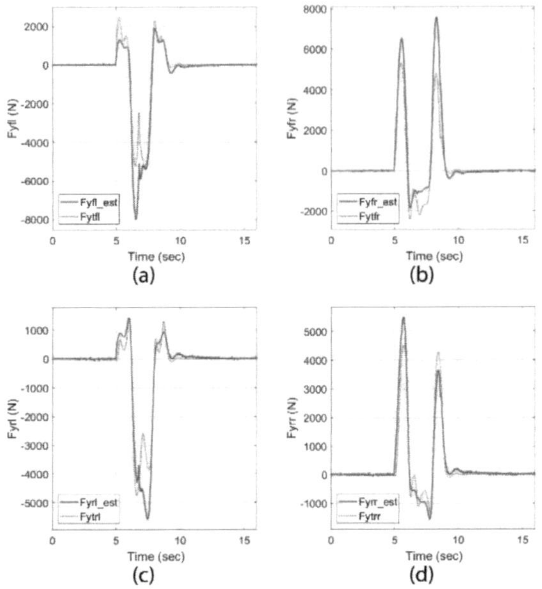

Figure 17. Estimated lateral forces in DLC manoeuvre (a) front-left (b) front-right (c) rear-left (d) rear-right. Fytij denotes actual tire force, while Fyij_est denotes estimated tire force.

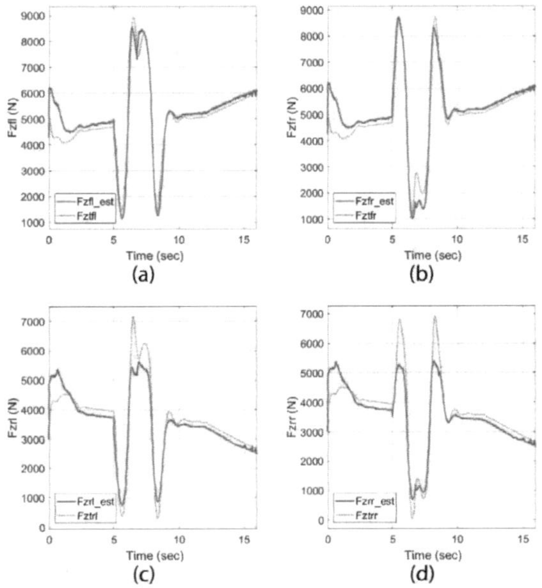

Figure 18. Estimated vertical forces in DLC manoeuvre (a) front-left (b) front-right (c) rear-left (d) rear-right. Fztij denotes actual tire force, while Fzij_est denotes estimated tire force.

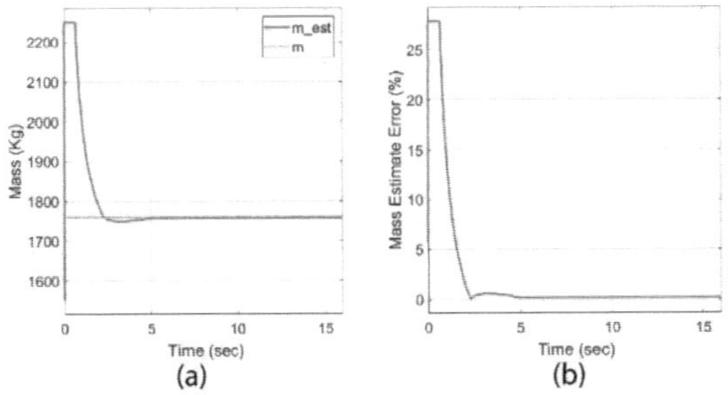

Figure 19. Mass estimation results for DLC manoeuvre. (a) comparison between actual mass m and estimated mass m_est. (b) percentage error for mass estimation.

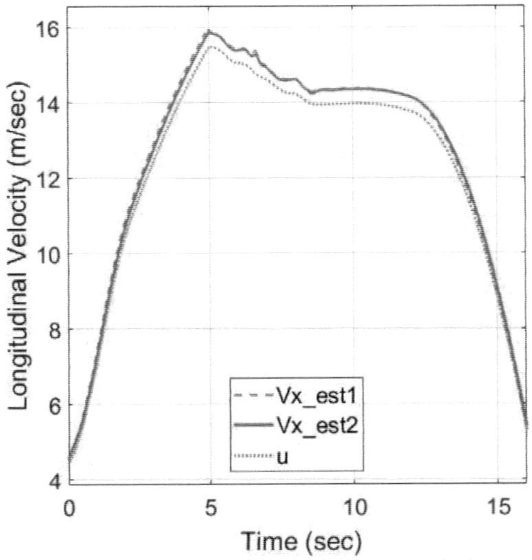

Figure 20. Longitudinal velocity estimation results in DLC manoeuvre. u denotes actual vehicle velocity, Vx_est1 denotes velocity estimation without slip estimation, Vx_est2 denotes velocity estimation with slip estimation.

Table 2. NRMS and maximum values for all modules outputs in DLC manoeuvre.

		FL	FR	RL	RR	N/A
F_x	NRMS	3.87	3.81	2.14	2.09	
	Max.	1728	1738	1683	1683	
F_y	NRMS	10.83	11.96	9.11	6.91	
	Max.	5229	5331	4756	4506	
F_z	NRMS	4.72	5.1	6.22	6.76	
	Max.	8955	8740	7167	6926	
V_x	NRMS					2.12
	Max.					15.47
m	PE					0.1335

4 Conclusion

The presence of active safety systems, such as Antilock Braking, Traction Control and Stability Control systems, have become increasingly important. The performance of such control systems depends on knowledge of vehicle dynamic parameters which are difficult to measure directly due to economic or technical reasons.

In [14] a novel unified estimation strategy was proposed for estimating tire forces in all 3 directions. The estimation strategy was of modular structure and has the advantage of being robust to vehicle mass change, since it includes mass estimation, and relies on no tire model, as tire models are prone to uncertainty due to tire aging or road surface condition.

Motivated by the work in [14] a modular estimation strategy for dynamic parameters pertaining to motion control applications is developed and presented, based on previous modules from [14] and 2 developed modules for vehicle mass estimation and longitudinal velocity estimation.

A 14-DOF vehicle model was simulated on Matlab in 3 manoeuvres, namely, Slalom manoeuvre on dry road, Slalom manoeuvre on wet road with additional 200Kg of mass, and Double-Lane-Change manoeuvre on a dry road. Simulation results shows good performance for the newly developed modules.

The new developed mass estimation module shows good performance while being simpler, relying only on minimal instrumentation.

Even though the new longitudinal velocity estimation relied on a modal tire model, it maintained an accurate output on varying road condition. The velocity estimate could be useful for motion control applications such as Active Cruise Control.

References

1. W. Cho, J. Yoon, S. Yim, B. Koo, K. Yi, "Estimation of Tire Forces for Application to Vehicle Stability Control", *IEEE Trans. Veh. Technol.*, **59**, 2, (2010).
2. M. Doumiati, A. Victorino, D. Lechner, G. Baffet, A. Charara, "Observers for vehicle tyre/road forces estimation: experimental validation", *Veh. Syst. Dyn.*, **48**, 11, (2010).
3. M. Doumiati, A. Victorino, A. Charara, D. Lechner, "Lateral load transfer and normal forces estimation for vehicle safety: experimental test", *Veh. Syst. Dyn.*, **47**, 12, (2009).
4. A. Rabhi, N.K. M'Sirdi, A. Elhajjaji, "Estimation of Contact Forces and Tire Road Friction", *Proc. Mediterranean Conf. Control Autom.*, 1-6, (2007).
5. S. Antonov, A. Fehn, A. Kugi, "Unscented Kalman filter for vehicle state estimation", *Veh. Syst. Dyn.*, **49**, 9, (2011).
6. J. Kim, H. Lee, S. Choi, "A robust road bank angle estimation based on a proportional–integral H∞ filter", *Proc. Inst. Mech. Eng. D. J. Autom. Eng.*, **226**, 6, (2012).
7. R. Ghandour, A. Victorino, M. Doumiati, A. Charara, "Tire/road friction coefficient estimation applied to road safety", *Proc. 18th Mediterranean Conf. Control Autom.*, 1485-1490, (2010).
8. H.B. Pacejka, I. Besselink, *Tire and Vehicle Dynamics*, (Elsevier, Butterworth-Heineman 2012).
9. R. Rajamani, D. Piyabongkarn, J.Y. Lew, J.A. Grogg, "Algorithms for Real-Time Estimation of Individual

Wheel Tire-Road Friction Coefficients", *Proc. Amer. Control Conf.*, 4682-4687, (2006).

10. B. Jaballah, N.K. M'Sirdink, A. Naamane, H. Messaoud, "Estimation of vehicle longitudinal tire force with FOSMO & SOSMO", *IJ-STA*, **5**, 1, (2011).
11. H. Imine, N.K. M'Sirdi, Y. Delanne, "Adaptive Observers and Estimation of the Road Profile", *Soc. Automotive Eng. World Cong.*, (2003).
12. L.R. Ray, "Nonlinear State and Tire Force Estimation for Advanced Vehicle Control", *IEEE Trans. Control Syst. Technol.*, **3**, 1, (1995).
13. K. Huh, C. Seo, J. Kim, D. Hong, "Active Steering Control Based on The Estimated Tire Forces", *Proc. Amer. Control Conf.*, 729-733, (1999).
14. A. Rezaeian, R. Zarringhalam, S. Fallah, W. Melek, A. Khajepour, S.K. Chen, N. Moshchuck, B. Litkouhi, "Novel Tire Force Estimation Strategy for Real-Time Implementation on Vehicle Applications", *IEEE Trans. Veh. Technol.*, **64**, 6, (2015).
15. M. Doumiati, A. Victorino, A. Charara, D. Lechner, "Virtual sensors, application to vehicle tire-road normal forces for road safety", *Proc. Amer. Control Conf.*, 3337-3343, (2009).
16. H.S. Bae, J. Ryu, J.C. Gerdes, "Road Grade and Vehicle Parameter Estimation for Longitudinal Control Using GPS", *IEEE Proc. Intel. Transp. Syst. Conf.*, (2001).
17. J. Grieser, "Method for Determining an Estimate of the Mass of a Motor Vehicle", *U.S. Patent*, 6,980,900, (2005).
18. A. Vahidi, A. Stefanopoulou, H. Peng, "Experiments for Online Estimation of Heavy Vehicle's Mass and Time-Varying Road Grade", *Proc. ASME IMECE*, 451-458, (2003).
19. A. Vahidi, M. Druzhinina, A. Stefanopoulou, H. Peng, "Simultaneous Mass and Time-Varying Grade Estimation for Heavy-Duty Vehicles", *Proc. Amer. Control Conf.*, 4951-4956, (2003).
20. V. Winstead, I.V. Kolmanovsky, "Estimation of Road Grade and Vehicle Mass via Model Predictive Control", *Proc. IEEE Conf. Control App.*, 1588-1593, (2005).
21. M.A. Wilkin, W.J. Manning, D.A. Crolla, M.C. Levesley, "Use of an extended Kalman filter as a robust tyre force estimator", *Veh. Syst. Dyn.*, **44**, 50-59, (2006).
22. J. Stephant, A. Charara, D. Meizel, "Virtual Sensor: Application to Vehicle Sideslip Angle and Transversal Forces", *IEEE Trans. Ind. Electron.*, **51**, 2, (2004).
23. A. Rehm, "Estimation of Vehicle Roll Angle", *Proc. 4th ISCCSP*, 1-5, (2010).
24. H.K. Fathy, D. Kang, J.L. Stein, "Online Vehicle Mass Estimation Using Recursive Least Squares and Supervisory Data Extraction", *Proc. Amer. Control Conf.*, 1842-1848, (2008).
25. Y. Zhang, H. Zhao, L. Yuan, H. Chen, "Slip Ratio Estimation for Electric Vehicle with In-wheel Motors based on EKF without Detection of Vehicle Velocity", *Proc. 28th CCDC*, 4427-4432, (2016).
26. S. Julier, J. Uhlmann, H.F. Durrant-Whyte, "A New Method for the Nonlinear Transformation of Means and Covariances in Filters and Estimators", *IEEE Trans. Autom. Control*, **45**, 3, (2000).
27. T. Shim, C. Ghike, "Understanding the limitations of different vehicle models for roll dynamics studies", *Veh. Syst. Dyn.*, **45**, 3, (2007).

Study on the Mechanism of Cold and Negative Temperature Coefficient in Natural Process of Gasoline Hydrocarbon

Qu Yakun , Long Jun , Zhou Han

Research Institute of Petroleum Processing,SINOPEC, 100083 Beijing,China

Email: yakun2015@126.com,longjun2002@126.com

Abstract. In this paper, the stoichiometric mechanism of gas phase oxidation process of gasoline hydrocarbons was studied through using theoretical stoichiometry. The reason of the phenomenon of cold flame and negative temperature coefficient in the reaction of hydrocarbon molecules before the flame was explained from the molecular level. During the gas phase oxidation process, the alkoxy radical RO· reacts with hydroxyl ·OH to form a relatively stable intermediate such as aldehyde (or ketone) and H_2O molecules, and the free radical chain reaction process.The temperature of the reaction process is very low, while the release of a large number of heat, the formation of aldehydes (or ketones) from the excited state back to the ground state when the emission of about 400nm wavelength of light blue fluorescence.

1 Introduction

According to the working principle of the ignition engine, one of the important ways to improve the engine power and reduce the fuel consumption is to improve the engine compression ratio. But with the increase of the engine compression ratio, in terms of fuel, it should be noted that the tendency of the fuel to generate knocking is increased simultaneous [1]. Therefore, the knocking phenomenon became a huge obstacle in the development and use of ignition engines in recent years. The knocking phenomenon in the ignition engine is the abnormal combustion caused by spontaneous combustion of unburned mixture when the normal flame has not yet reached, which is closely related to the high phase oxidation characteristics of the hydrocarbons. From the point of view of oxidation kinetics, the vapor-phase oxidation of hydrocarbons is often characterized by cold flame, that is, weak glow in the phase of gas phase oxidation of hydrocarbon molecules, and the negative temperature coefficient will be occured, that is, the higher the initial temperature, the ignition delay time will be longer, as shown in Figure1. These phenomenons are one of the important characteristics of hydrocarbon gas phase oxidation. According to the experimental observation of many researchers, in addition to methane, methanol, formaldehyde, ethylene and benzene a few hydrocarbon molecules, almost all air heating oxidation of the hydrocarbons will produces cold flame and NCT phenomenon. And the temperature of cold flame produced by the spontaneous combustion process of hydrocarbon molecules has the following regularity: With the increase of carbon atoms in the hydrocarbon molecules, the temperature of the cold flame is significantly reduced; the olefin, naphthenes and aromatic hydrocarbons produce relatively high temperature cold flame, which is very similar to the tendency of generate knocking in the ignition engine.

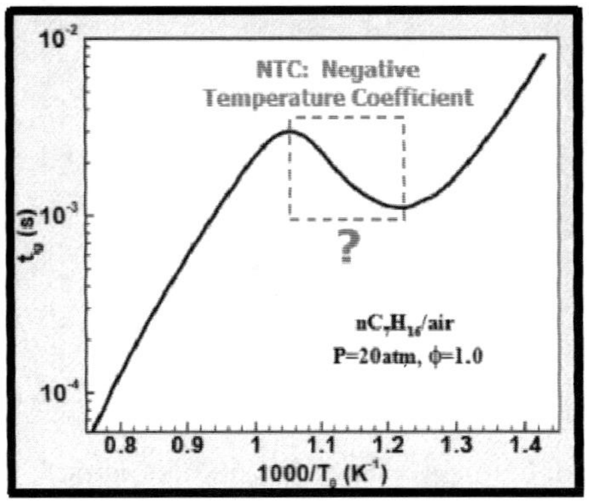

Figure 1. Negative temperature coefficient (NTC) in the process of hydrocarbon spontaneous combustion

With the rapid development of high-precision analysis and characterization technology, the current understanding of finished gasoline has risen to the molecular level, the components of finished gasoline is more than 300 kinds [2]. The range of hydrocarbon molecules is from 4 to 12, and the temperature range is from 30°C to 220°C, including straight and isoparaffins, naphthenes, olefins, aromatic hydrocarbons and oxygen-containing agents[3]. It is the core content of the basic research of combustion physics to study the various elementary reactions and reaction rates occurring at the molecular level of gasoline hydrocarbons [4]. However, recent research works are focused on the in-cylinder physical processes, such as the improvement and improvement of spray, evaporation, and turbulence models, and the importance of fuel chemistry are ignored. Molecular details of chemical composition such as fuel composition and molecular structure, depth understanding combustion mechanism are also lacked, thus the feasibility of controlling combustion and optimizing the concept of combustion from a chemical reaction point of view are limited. Based on the above background, in the paper, it adopts the theoretical chemistry method to study the pre-flame reaction mechanism of the gas phase oxidation process of gasoline C_8 hydrocarbon molecules, find out the reason of cold flame and negative temperature coefficient in the reaction of hydrocarbon molecules before the molecular level. It has great important significance on essence of the cause of knocking, the oil refining and engine design.

2 Calculation Method and Model

Due to the complexity and volatility of the actual fuels, commercial fuels (gasoline, diesel, kerosene) contain hundreds or more hydrocarbon mixtures. Although advanced chemical analysis can express quantification of the constituent information, and it is difficult to accurately characterize the actual fuel [5]. In order to simulate the physical and chemical properties of the actual fuel, it selected the C_8 alkane octadecane octane number of which is -17. And the initial molecular model of the monomer hydrocarbon is established through referring to the molecular structure parameters in the NIST calculation and the reference database (http//cccbdb.nist.Gov). The structural information is as shown in Figure 2.

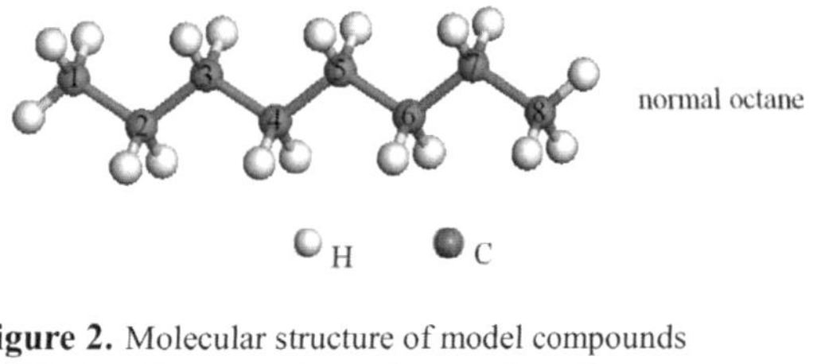

Figure 2. Molecular structure of model compounds

Under the premise of ensuring the accuracy of calculation, the calculation efficiency is selected, and the DFT based on the quantum chemical calculation method is selected. The hybrid function B3LYP [6, 7], the wave function is developed in the form of triplet plus polar group TNP (Triple Numerical plus polarization), and the electron treatment is all electronically.

The transitional state search, geometric optimization, and other electronic properties are calculated through using the $DMol^3$ module in the BIOVIA Materials Studio 2017 R2 software package [8]. Fragment analysis, energy decomposition, Hirshfeld charge analysis of the shell system are based on the ADF module [9, 10]. And the calculation of high-spin state of the open shell adopts BAND module[11-14].

In the transition state search, the Linear Synchronous transient and Quadratic Synchronous Transit (LST/QST) method are used and the RMS convergence criterion is 0.002Ha/Å. The structural optimization convergence criterion is 1×10^{-7}Ha, the maximum force is 1×10^{-6}Ha/Å, the maximum displacement is 1×10^{-6}Å, and the frequency analysis shows no virtual frequency.

The long-range interaction of DFT+D is calculated by TS method. The convergence of self-consistent field SCF (self-consistent field) is accelerated by DIIS method. The multi-pole expansion of density is 16-pole expansion, SCF convergence criterion is 1×10^{-6}Ha.

The two fragments of the interaction were analyzed by using the NOCV (Natural orbitals for chemical valence) method and the ETS-NOCV (extended transition state-natural orbitals for chemical valence) method under the Kohn-Sham molecular orbital theory model.

3 Results and Discussion

According to the author's previous study on the initiation process of hydrocarbon molecular gas oxidation chain, it is known that the hydrocarbon molecule is inserted into the C-H bond through O_2 to produce the peroxidase ROOH of the hydrocarbon.

Then the o-o lysis of peroxide ROOH causes the hydroxyl radical RO- and hydroxyl - OH to be generated, as shown in equation (1).

O2 insert C-H key::

$$RH + O2 \rightarrow ROOH \rightarrow RO\cdot + \cdot OH \qquad (1)$$

In the case of n-octane, the alkoxy radical $C_8H_{17}O$ and the hydroxyl·OH will be produced by the chain initiation process. As shown in Fig.3, the generated alkoxy radical $C_8H_{17}O$ can produce β cracking at a high temperature, but the reaction activation energy is high, in the hydrocarbon molecular combustion process of the flame before the reaction phase, the temperature is low, such reactions are difficult to occur. At the same time, the formation of the alkoxy radical $C_8H_{17}O$ and hydroxyl OH can occur between the H-reaction, but also with another alkane occurred in the H-reaction. The former product is a more stable compound, which in turn produces a new free radical.

It is well known that the negative temperature coefficient phenomenon occurs during the spontaneous combustion of hydrocarbon molecules which is related to the degradation of the branched chain reaction of the hydrocarbon molecules in the free radical chain reaction, that is, the temporary chain termination characteristic occurs during the reaction. The reaction between the alkoxy radical $C_8H_{17}O$·and the hydroxyl OH produced by the chain initiation has a temporary chain termination property, and the radical reaction can also be considered to belong to the degraded branched chain reaction. The

 https://doi.org/10.1051/matecconf/201816602007

activation energy of this reaction is about 0 kJ·mol^{-1}, the enthalpy becomes -439.413 kJ·mol^{-1}, and the activation energy of the reaction is very low, and the reaction is very easy to be studied by quantum chemical calculation based on DFT method. The reaction has characteristic of strong exothermic, so from thermodynamical view is also very easy to be happened.

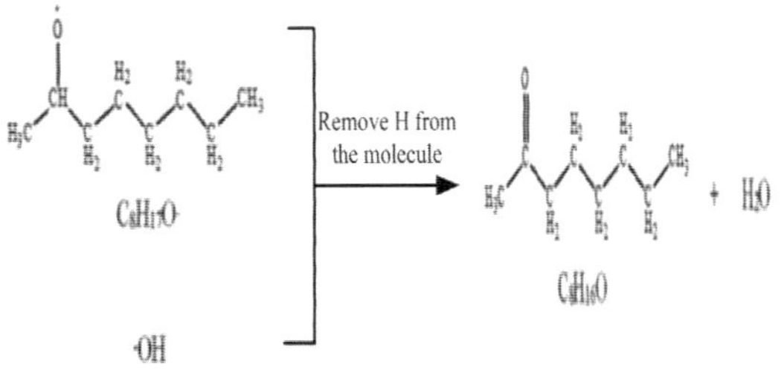

Figure 3. The reaction between free radicals generated by chain initiation

The change of the orbital component of the reaction process is studied based on the molecular orbital theory, as shown in Fig. 4 and Fig. 5. The single occupancy orbital (SMO) of the hydroxyl radicals in the hydroxyl OH is the 2px orbital of the·O·atoms in the hydroxyl OH, which is -8.547eV after the formation of the·OH and in the presence of the oxygen radical $C_8H_{17}O$ (Lowest Unoccupied Molecular Orbital, LUMO) during the reaction which occupied Molecular Orbital, HOMO).

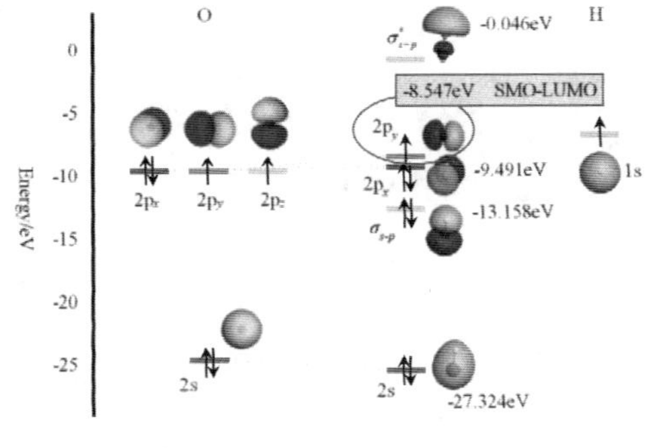

Figure 4. Orbital residence of hydroxyl OH

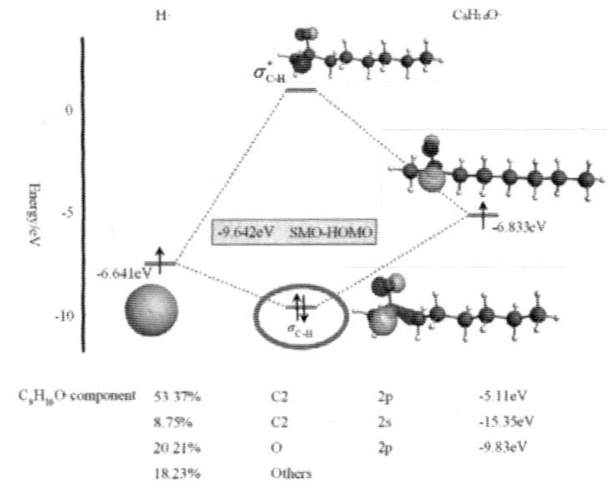

Figure 5. C-H bond NOCV in the presence of hydroxyl radical $C_8H_{17}O$ with hydroxyl OH radicals

The C-H bond in the $C_8H_{17}O$·participant in the reaction is the σ bond formed by the linear combination of the orbital of the H atom and the 1s orbital of the H atom, and the formed σ C-H bonding orbit is affected by the free radical. And the CH bond is weakened, so that the reaction energy barrier of this reaction is very low. In this paper, the π-C bond is doped into the 2p orbital component of the O atom.

During the reaction, electrons are transferred from the σC-H bond orbital of the alkoxy radical $C_8H_{17}O$·, the highest occupied molecular orbital (HOMO) of the alkoxy radical $C_8H_{17}O$· to the LUMO of hydroxyl ·OH.

The reaction of 2-octanone and H_2O with the reaction of hydroxyl radical $C_8H_{17}O$ with hydroxyl·OH is a relatively stable substance which produces a similar chain termination effect during the vapor phase oxidation of the hydrocarbon molecule. Therefore, it is considered that the reaction is a hydrocarbon molecule The mechanism of NTC phenomenon in the gas phase oxidation process is as shown in Fig.6, which compares the mechanism of the occurrence of NTC phenomenon in the gas phase oxidation process of hydrocarbon molecules and the mechanism found in the study.

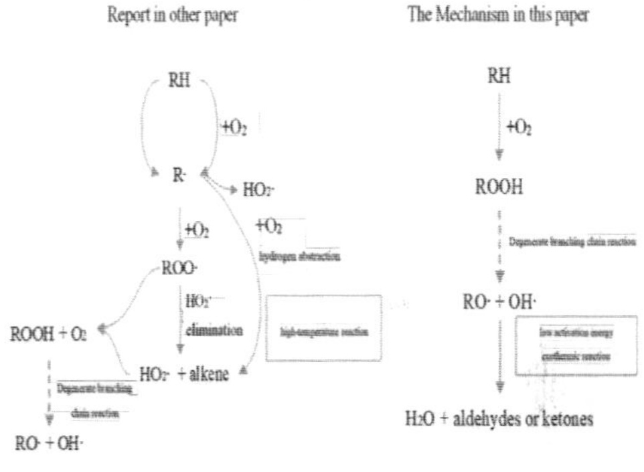

Figure 6. Different mechanisms of NTC phenomenon

The mechanism reported in published literature cannot explain the relationship between the cold flame and the NTC phenomenon comprehensively and cannot explain why the cold flame is produced at the same time, why there is water vapor, which is not consistent with the experimental phenomenon in some parts. In this study, the proposed mechanism of the reaction is not only tan make up for the above problems, but also can find that methane, benzene and other molecules without NTC phenomenon is due to the hydroxyl OH process, the it first to capture with the oxygen free radicals and O atoms connected to the C atom on the H, methane, benzene and other molecular chains caused by the production of alkoxy and O atoms connected to the C bond formed by the bond with a strong bond, hydroxyl·OH reaction activation can be very large, it is difficult to occur this reaction, so the gas phase oxidation process is not cold flame and NTC phenomenon. it explains why there are no cold flame molecules and no NTC phenomenon from the perspective of the reaction mechanism.

In the processes of Alkoxy radicals $C_8H_{17}O$·and hydroxyl ·OH,it can produce some aldehydes and ketones and emit large amounts of heat. Therefore, it is presumed that the glow of the cold flame emitted during the gas phase oxidation of the hydrocarbon molecules is likely to be the oxygen radical $C_8H_{17}O$ and in order to further determine the correctness of the reaction mechanism, the hydrogenation of aldehydes or ketones which may be produced by a series of hydrocarbon gas, phase oxidation processes is calculated by the reaction of hydroxyl ·OH to produce aldehydes or ketones excited from the excited state to the ground state. Spectrum is as shown in Fig. 7.

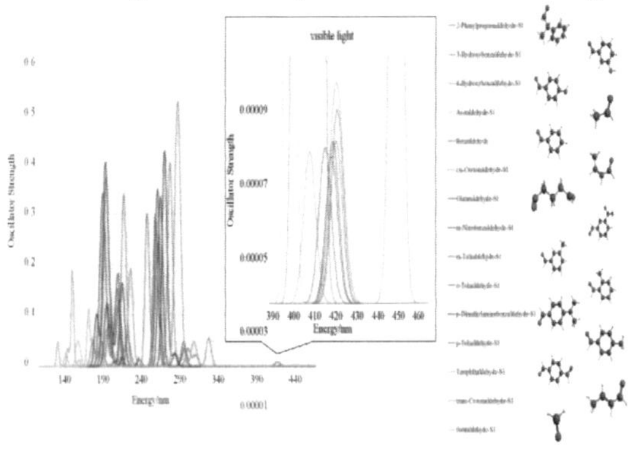

Figure 7. Some aldehydes, ketones and other molecules of the UV absorbable spectrum

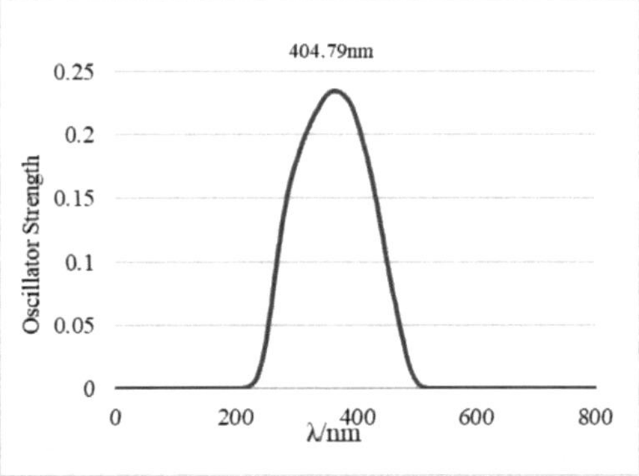

Figure 8. Fluorescence emission spectra of single first excited states of 2-octanone

Generally human eyes can perceive the wavelength of electromagnetic waves between 380~780nm, the purple visible wavelength range is about 380~420nm, blue visible wavelength is about 440~475nm. During the combustion process, Ultraviolet absorption light wavelength of the partial absorption of aldehydes, ketones and other oxides containing the first single excited state of the oxide is about 390 ~ 460nm, which is basically coincident with the blue-violet wavelengths perceived by the human eye. The glow color released by the cold flame observed by the experiment is also basically coincident. In order to study the chemical nature of the cold flame during the gas phase oxidation of hydrocarbon molecules, the fluorescence emission spectrum of the single first excited state of 2-octanone is calculated. As shown in Fig. 8, the peak wavelength is 404.79 nm, and the excitation energy of the single first excited state of 2-octanone is 295.52 kJ· mol^{-1}, which is much smaller than that of the reaction. It is further proved that the cold flame produced by the gas phase oxidation of hydrocarbon molecules is the free radicals in the free radical reaction process, the corresponding aldehydes (or ketones) generated by the reaction of the hydroxyl ·OH are excited by the large amount of heat emitted by the reaction, the resulting single first excited state returns to ground state will emit blue violet fluorescence.

4 Conclusion

In the process of vapor phase oxidation of hydrocarbons, the oxygen radical RO·and hydroxyl·OH are generated by chain initiation, and the reaction between two reactive radicals will form the corresponding derivatives of aldehydes, ketones or other hydrocarbons and H_2O molecules, the energy barrier of the reaction process is very low, and will approach 0, and a large amount of heat is released, so that the generated aldehyde or ketone is excited from the ground state to the single first excited state, and when it returns from the excited state to the ground state, it will emits about 400nm blue-violet fluorescence.

In the process of gas phase oxidation of hydrocarbons, in the progress of the reaction, the temperature will be risen, the reaction will be slow, the main reason is the chain reaction lead to produce oxygen radical RO· and hydroxyl ·OH and to produce a corresponding halide, ketone or other hydrocarbon derivatives and H_2O molecules and other relatively stable intermediates, which cause vapor phase oxidation of hydrocarbons free radical chain reaction process temporarily terminated.

References

1. Wang Ning and Wen Hao, the chemical composition and structure of gasoline fuel and the relationship between the structure and anti-explosive properties of the oil refining and chemical industry, 27 (2),41-46(1996)
2. Liu Yuhua, Ma Wenyou, Full Analysis of Monomer Hydrocarbons in Catalytic Cracking Gasoline. Journal of Qiqihar University (Natural Science Edition), (2),68-69(1999)
3. Allison, T.C. Workshop on Combustion Simulation Databases for Real Transportation Fuels. In The Workshop on Combustion Simulation Databases for Real Transportation Fuels,(2003)
4. Liu Zelong. Explore the oil molecules "mettle." Sinopec, (2),50-51(2014)
5. Xiao Gan, Zhang Yusheng, Jiang Guangjun. Systematic Construction and Verification of Gasoline Multi-component Characterization of Fuel Simplification Dynamics Model. Journal of Physical Chemistry, (4), 879-892(2016)

6. Becon, A., Density functional thermochemistry. III. The role of exact exchange. J. Chem. Phys, 98, 5648-5652(1993)
7. Stephens, Pauk, Ab initio calculation of vibrational absorption and circular dichroism spectra using density functional force fields. The Journal of Physical Chemistry, 98 (45),11623-11627(1994)
8. Delley, B., DMOL is a density functional theory (DFT) program distributed by Accelrys, Inc. J. Chem. Phys, 92(1990)
9. Te Velde, G.t., Chemistry with ADF. Journal of Computational Chemistry, 22 (9), 931-967(2001)
10. Guerra, C.F., Towards an order-N DFT method. Theoretical Chemistry Accounts,99(6), 391-403(1998)
11. Te Velde, G. and E. Baerends, Precise density-functional method for periodic structures. Physical Review B, 44 (15), 7888(1991).
12. Wiesenekker, G. and E. Baerends, Quadratic integration over the three-dimensional Brillouin zone. Journal of Physics: Condensed Matter, 3 (35), 6721(1991).
13. Franchini, M., P.H.T. Philipsen, and L. Visscher, The Beckm fuzzy cells integration scheme in the Amsterdam Density Functional program suite. Journal of computational chemistry,34(21), 1819-1827(2013).
14. Franchini, M., Accurate Coulomb Potentials for Periodic and Molecular Systems through Density Fitting. Journal of Chemical Theory and Computation, 10 (5), 1994-2004(2014).

Road Bank Angle Estimation for Three Wheel Tilting Vehicle Using Multi Model Estimator

Kicheol Jeong[1], Seibum Choi[1,a]

[1]KAIST, Department of Mechanical Engineering, 34141 Deajeon, Republic of Korea

Abstract. In recent years, the need for micro-mobility, especially three-wheel vehicles, is increasing to address pollution and traffic congestion problems. With regard to the development of tilting vehicles, the precise tilt angle is important information in the tilting mechanism. Since the road environment affects the vehicle tilt angle, the road bank and hill angle have to be estimated to optimize the tilt control system. Furthermore, a new tilt mechanism is required due to the energy consumption of the previous active tilting mechanism. This paper introduces the road state observer. In this paper, the observer that combines a kinematic model with a dynamic model of a three wheel vehicle is proposed. The dynamic model has four states, including lateral velocity, yaw rate, tilt angle, and tilt angle ratio. Similarly, kinematic model has two states, including roll and pitch angles. It is assumed that the data set received from the six-dimensional inertial measurement unit including the vehicle acceleration and angular velocity of all axes is available. To verify the proposed algorithm, simulation verification using Carsim ADAMS and Matlab&Simulink is performed and a discussion of the result is provided. In addition, this paper proposes a semi-active tilt system.

1 Introduction

Recently, conventional transportation systems cause many problems. From an ecological point of view, interest in air pollution and depletion of fossil fuel problems has increased. Furthermore, in modern cities, traffic volume and parking lot shortages are becoming more serious

In this situation, narrow and short micro-mobility, especially three wheel vehicle, is interested in solving this phenomenon [1, 2]. This small vehicle operates as an electric power train system, so there is no exhaust, parking and traffic congestion. However, because of the body shape of this vehicle, it has a high center of gravity. This physical property makes this vehicle vulnerable to rollover. To overcome this drawback, a three wheel vehicle must include a lean system that compensates for the roll stability of the mobility during cornering maneuver [3-5].

It is important to determine the exact tilt angle to control the vehicle tilt system. Because road bank and hill angles have an impact on the vehicle tilt angle, the road characteristics have to be estimated to optimize the tilting control system. In previous studies on road angles, S. Kidane directly calculated road bank angles using vehicle dynamic models [6]. However, this method has a limitation that the logic is unstable when the vehicle has a small tilt angle.

In this paper, the road state observer combining the three wheel vehicle dynamic model and the kinematic model is proposed. The dynamic model has four states, including lateral velocity, yaw rate, tilt angle and tilt angle rate. Likewise, the kinematic model has two states including roll and pitch angle. It is assumed that a set of data received from a six dimensional-inertial measurement unit (6D-IMU) that provides the vehicle linear acceleration and angular rates of all axes is available.

The proposed observer in this paper is organized as follows. First, the pseudo-measurement is generated from a multi-model based pseudo-angle generator. This generator is based on the three wheel vehicle dynamic model and a Kalman filter based on type 1 Euler angle kinematic model. Depending on the condition of the vehicle state, the generator generates an accurate pseudo-measurement that is used to estimate the vehicle roll and pitch angle. Next, the vehicle roll and pitch angle are obtained with a kinematic model-based linear observer. This observer is based on a kinematic model and uses pseudo measurement as the sensor measurement. Finally, the bank angle is obtained from the road bank angle calculation process.

This paper is organized as follows. Section 2.1 contains the demonstration of the three wheel vehicle dynamics. Section 2.2 explains the pseudo-measurement generator. Section 2.3 explains linear estimator, section 2.4 explains nonlinear estimator and section 2.5 describe the bank angle calculation scheme. Next, section 3 shows the simulation verification result of the observer. Finally, section 4 deals with the conclusion of this paper.

[a] Corresponding author: sbchoi@kaist.ac.kr

2 Road bank estimator

2.1 Three wheel vehicle dynamic model

2.1.1 Vehicle dynamic model using gravity transformation

To illustrate the tilt behavior of a vehicle, a specialized three wheel vehicle dynamic model is required. In particular, the road angle affects the vehicle tilt angle. Therefore, when deriving a vehicle dynamic model, the angle of the road should be considered.

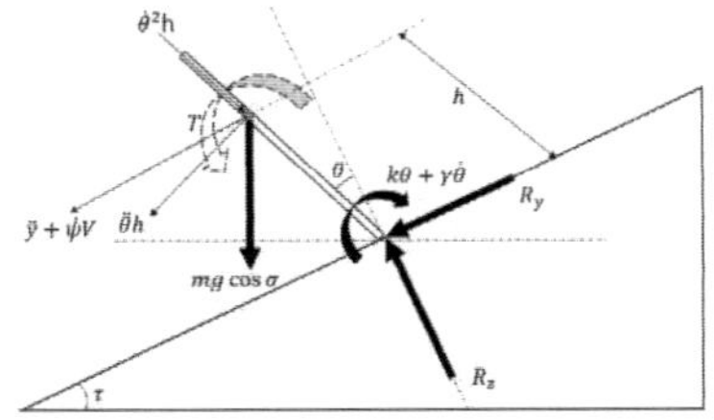

Figure 1. Three wheel vehicle free body diagram

As shown in Figure 1, the road angle has a significant impact on vehicle tilt dynamics. According to this free body diagram, the gravity coordinate transformation matrix can be obtained.

$$R_S = \begin{bmatrix} \cos\sigma & 0 & \sin\sigma \\ 0 & 1 & 0 \\ -\sin\sigma & 0 & \cos\sigma \end{bmatrix} \tag{1}$$

$$R_B = \begin{bmatrix} 1 & 0 & 0 \\ 0 & \cos\tau & -\sin\tau \\ 0 & \sin\tau & \cos\tau \end{bmatrix} \tag{2}$$

$$G_T = R_B R_S G_N = \begin{bmatrix} -g\sin\sigma \\ g\sin\tau\cos\sigma \\ -g\cos\tau\cos\sigma \end{bmatrix} \tag{3}$$

The above matrix is a gravity transformation matrix. Equation (1) is a longitudinal gradient conversion matrix, and equation (2) is a lateral gradient conversion matrix. The matrix makes deriving mathematical model process simple. Using this transformation matrix, three-wheel vehicle dynamic model can be obtained as follows.

$$A = \begin{bmatrix} -\frac{\alpha}{mv}\left(2C_f + C_r\right) & -\left(v + \frac{\alpha}{mv}\left(2C_f l_f - C_r l_r\right)\right) & \frac{hk}{I_{xx}} - \frac{mh^2}{I_{xx}}g & \frac{h\gamma}{I_{xx}} \\ \frac{C_r l_r - 2C_f l_f}{vI_z} & -\frac{2C_f l_f^2 + C_r l_r^2}{vI_z} & 0 & 0 \\ 0 & 0 & 0 & 1 \\ h\frac{2C_f + C_r}{I_{xx}v} & h\frac{2l_f C_f - l_r C_r}{I_{xx}v} & \frac{mgh}{I_{xx}} - \frac{k}{I_{xx}} & -\frac{\gamma}{I_{xx}} \end{bmatrix} \tag{4}$$

This is a state space representation model, which includes the lateral velocity, yaw rate, tilt angle, and tilt velocity. In this paper, vehicle tire force is assumed to be a linear tire force. In previous studies like [7], other vehicle tire model is presented, but are generally suitable for high speed conventional four wheel vehicles. This assumption is reasonable because a three wheel tilting vehicle has a relatively low speeds in general.

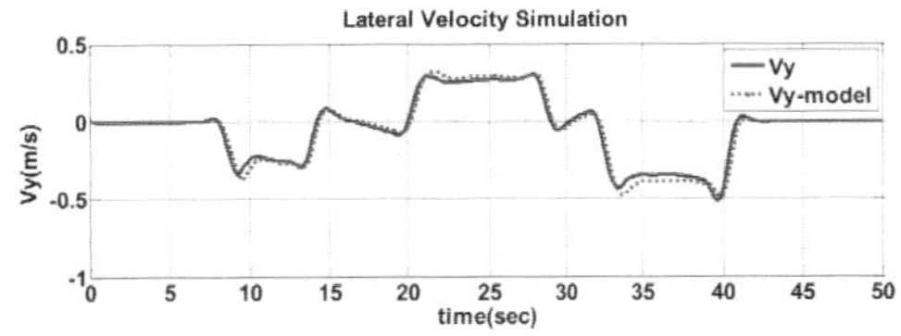

Figure 2. Model verification result

The above figure shows the model verification result. Model and simulation plant outputs for the same steering inputs are similar. Therefore, the model is well defined and can be used to construct estimators.

2.1.2 Design the desired tilt angle

In this paper, the lateral acceleration of the vehicle can be obtained from the sensor. However, the sensor signals are contaminated with roadways, vehicle inclination angles, and tilt angles.

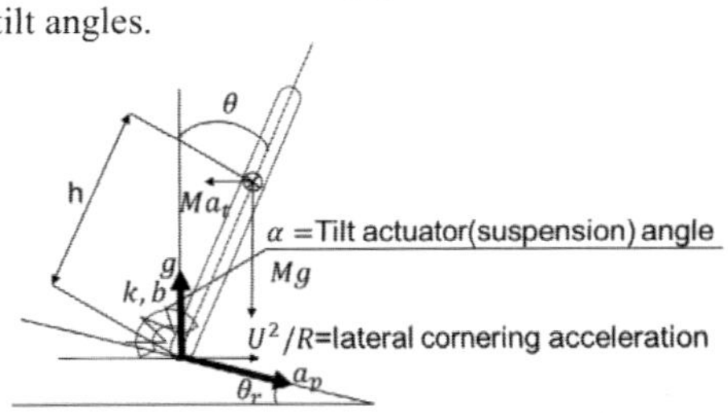

Figure 3. Earth fixed coordinate model

This figure shows the cornering with banked road situation. In this situation, the road bank, turning radius and vehicle velocity have an impact on the vehicle roll dynamics.

$$I\ddot{\theta} = mgh\sin\theta - mha_c + b\left(\dot{\alpha} + \dot{\theta}_r - \dot{\theta}\right)$$
$$+k(\alpha + \theta_r - \theta) \tag{5}$$

$$a_c = \frac{U_1^2}{R_0} \tag{6}$$

$$a_p = \frac{U^2}{R}\cos\theta - g\sin\theta \tag{7}$$

$$\tan\theta_d = \frac{U^2}{Rg} \tag{8}$$

This equation shows that the road bank angle affects the desired tilt angle. Equation (8) is the equation for the desired tilt angle. According to this equation, lateral acceleration does not affect vehicle movement when the vehicle is tilted to the desired tilt angle.

2.2 Pseudo measurement generator

2.2.1 Dynamic model based pseudo measurement generation

In this section, a pseudo measure generator is proposed. Since the vehicle roll and pitch angle cannot be measured, the pseudo measure replaces the actual measurement of the design roll and pitch angle observer. In a previous study, Eric Tseng introduced a pseudo measurement based on Euler angles of type 1 [8]. However, this study assumed that the lateral speed of the vehicle was negligible. Unlike other conventional vehicles, the three wheel tilting vehicle has a relatively high lateral velocity because tilt angular velocity is applied to the lateral velocity when the vehicle is tilted. Therefore, this paper introduces the lateral velocity compensation method using Kalman filter. The Kalman filter model is based on a dynamic model of a three-wheeled vehicle, which includes yaw rate, tilt angle and tilt speed. Then the vehicle lateral velocity can be obtained and compensate the kinematic model based pseudo measurement.

$$\begin{cases} \hat{x}_k^- = F_{k-1}\hat{x}_{k-1}^+ + G_{k-1}u_{k-1} \\ P_k^- = F_{k-1}P_{k-1}^+F_{k-1}^T + Q_{k-1} \\ K_k = P_k^- H_k^T\left[H_k P_k^- H_k^T + R_k\right]^{-1} \\ \hat{x}_k^+ = \hat{x}_k^- + K_k\left[z_k - H_k\hat{x}_k^-\right] \\ P_k^+ = (I - K_k H_k)P_k^- \end{cases} \tag{9}$$

$$\theta_{refM} = \sin^{-1}\left(\frac{\dot{v}_{wheel} - a_x - r\cdot\hat{v}_y}{g}\right) \tag{10}$$

$$\phi_{refM} = \sin^{-1}\left(\frac{-\dot{\hat{v}}_y + a_y - r\cdot v_{wheel}}{g\cdot\cos\theta_{refM}}\right) \tag{11}$$

With this dynamic model-based Kalman filter, relatively accurate pseudo-measurements can be obtained when the vehicle is in a steady state. However, when the vehicle is in a transient state, the accuracy of this dynamic model-based pseudo measurement decreases. Therefore, another pseudo measurement generation algorithm is needed to ensure accuracy in transient situations.

2.2.1 Kinematic model based pseudo measurement generation

The pseudo roll and pitch angle can be obtained kinematic model. The kinematic roll and pitch model are expressed as the following:

$$\phi_{\text{int}} = \int p + (q\sin\phi + r\cos\phi)\tan\theta dt \tag{12}$$

$$\theta_{\text{int}} = \int q\cos\phi - r\sin\phi dt \tag{13}$$

Then another pseudo measurements based on the kinematic model based Kalman filter can be obtained. This measurements is accurate under transient conditions, but relatively inaccurate when the vehicle is state.

According to equation (9), the high measurement error covariance makes the Kalman filter estimation result dependent on the state model rather than the measurement. By the contrary, the low measurement error covariance makes estimation result dependent on the measurement information. Therefore, in this paper, steady state factor (SF) that adjust the filter measurement error covariance is proposed. This element is a function of the norm of the acceleration. If the vehicle is in transient state, SF is reduced to zero. In the contrary, if the vehicle is stable, SF is increased to one.

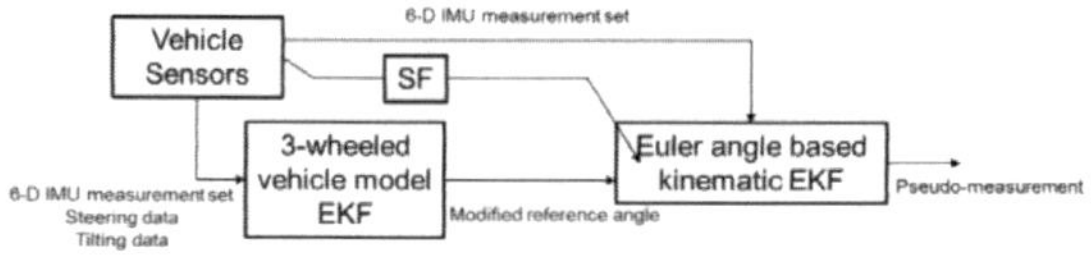

Figure 4. Pseudo-measurement generation scheme

2.3 Linear roll and pitch angle estimator

This section, linear vehicle roll and pitch angle estimator is developed. The linear estimator is based on simple kinematic model.

$$\dot{\phi} = p + (q\sin\phi + r\cos\phi)\tan\theta \tag{14}$$

$$\dot{\theta} = q\cos\phi - r\sin\phi \tag{15}$$

$$\Delta\phi = q\cdot\sin\phi\cdot\tan\theta + r\cdot(\cos\phi\cdot\tan\theta - \theta) \tag{16}$$

$$\Delta\theta = q\cdot(\cos\phi - 1) + r\cdot(\phi - \sin\phi) \tag{17}$$

The delta term above is a nonlinear element of equation (14) and (15). Even if the vehicle is leaning from the maximum tilt angle, the magnitude of this nonlinear term is negligible. Therefore, the linear estimator can be designed and the form of the estimator as follows.

$$\begin{bmatrix} \dot{\hat{\phi}} \\ \dot{\hat{\theta}} \end{bmatrix} = \begin{bmatrix} 0 & r \\ -r & 0 \end{bmatrix}\cdot\begin{bmatrix} \hat{\phi} \\ \hat{\theta} \end{bmatrix} + L\cdot\begin{bmatrix} \phi_{pseudo} - \hat{\phi} \\ \theta_{pseudo} - \hat{\theta} \end{bmatrix} \tag{18}$$

$$L = \begin{bmatrix} L_1 & L_2 \\ L_2 & L_3 \end{bmatrix} \tag{19}$$

$$\begin{bmatrix} \dot{\hat{\phi}} \\ \dot{\hat{\theta}} \end{bmatrix} = \begin{bmatrix} -L_1 & r - L_2 \\ -r - L_2 & -L_3 \end{bmatrix}\cdot\begin{bmatrix} \hat{\phi} \\ \hat{\theta} \end{bmatrix} + L\cdot\begin{bmatrix} \phi_{pseudo} \\ \theta_{pseudo} \end{bmatrix} \tag{20}$$

$$A^* = \begin{bmatrix} -L_1 & r - L_2 \\ -r - L_2 & -L_3 \end{bmatrix} \tag{21}$$

$$\left|\lambda I - A^*\right| = \left\|\begin{bmatrix} \lambda + L_1 & -r + L_2 \\ +r + L_2 & \lambda + L_3 \end{bmatrix}\right\| = (\lambda + L_1)(\lambda + L_3) - (L_2^2 - r^2) \tag{22}$$

Observer gain (19) should be selected to stabilize the observer system (21). According to equation (22), if the diagonal elements of the observer gain matrix are positive and the other elements are equal to yaw rate, then this observer is stable.

2.4 Nonlinear roll and pitch angle estimator

Therewith, in this paper, nonlinear estimation logic also proposed. Nonlinear terms are negligible, but large tilt angles make nonlinear effects larger. The nonlinear estimator can be established as the following:

$$\dot{\hat{\phi}} = p + (q\sin\hat{\phi} + r\cos\hat{\phi})\tan\hat{\theta} + k_1 \operatorname{sgn}(\phi_{pseudo} - \hat{\phi}) \quad (23)$$

$$\dot{\hat{\theta}} = q\cos\hat{\phi} - r\sin\hat{\phi} + k_2 \operatorname{sgn}(\theta_{pseudo} - \hat{\theta}) \quad (24)$$

$$\dot{s}_1 = \dot{\varepsilon}_{\phi} = \Delta f_{\phi} - k_1 \operatorname{sgn}(s_1) \quad (25)$$

$$\dot{s}_2 = \Delta f_{\theta} - k_2 \operatorname{sgn}(s_2) \quad (26)$$

In the worst case,

$$\Delta f_{\phi} = \alpha_{\phi} \operatorname{sgn}(s), \Delta f_{\theta} = \alpha_{\theta} \operatorname{sgn}(s) \quad (27)$$

Therefore,

$$s_1\dot{s}_1 = s_1[\alpha_{\phi} \operatorname{sgn}(s_1) - k_1 \operatorname{sgn}(s_1)] = -s_1\eta_1 \operatorname{sgn}(s_1) < 0 \quad (28)$$

$$s_2\dot{s}_2 = s_2[\alpha_{\theta} \operatorname{sgn}(s_2) - k_2 \operatorname{sgn}(s_2)] = -s_2\eta_2 \operatorname{sgn}(s_2) < 0 \quad (29)$$

$$k_1 \equiv \alpha_{\phi} + \eta_1 \quad (30)$$

$$k_2 \equiv \alpha_{\theta} + \eta_2 \quad (31)$$

2.5 Road bank angle calculation

Road bank angle calculation is a difficult subject of vehicle roll dynamics. It is difficult to estimate the road bank angle without GPS / GNS because the road bank angle and suspension displacement simultaneously contaminate the vehicle sensor signals. However, the three wheel tilting vehicle can identify the angle due to the suspension deflection due to the presence of the tilt angle sensor. Therefore, the road bank angle is simply calculated as the difference between the vehicle total roll angle and the tilt angle

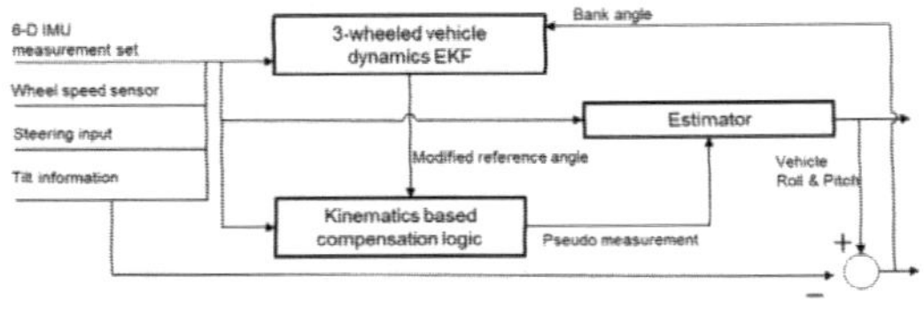

Figure 5. Road bank angle estimation schem

3 Simulation Result

To verify the proposed algorithm, simulation verification using Carsim and Matlab&Simulink is conducted. Various simulations are performed to test the robustness and stability of the algorithm. The table below enumerates various maneuver to test proposed algorithm.

Table 1. Test maneuver

	Road	Velocity	Behavior
Case 1	15 deg ramp bank	80kph	Straight+DLC
Case 2	Normal road	80kph	Normal driving
Case 3	5deg ramp bank	80kph	Fishhook
Case 4	Flat	20kph	Sine wave Steer
Case 5	Flat	40kph	Sine wave steer with tilting control

Case 1 maneuver is performed to verify the performance of the estimator for large amounts of lateral velocity and road bank angles. It is expected that there is a difference between the proposed estimator performance and previous estimates that ignore vehicle lateral velocity. On the other hand, Case 2 is conducted to evaluate the normal road performance of the proposed estimator. Case 3 is called "Fishhook", which is a test of vehicle stability testing. This extreme situation is simulated to evaluate the estimator stability and performance. It can be seen that the Case 4 and Case 5 simulation are not performed on the banked road. A sinusoidal steering input is entered to evaluate performance on the zero degree road bank angle.

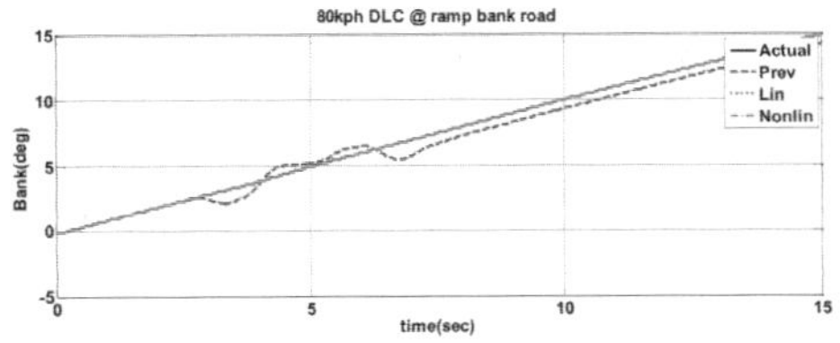

Figure 6. Case 1 Road bank angle

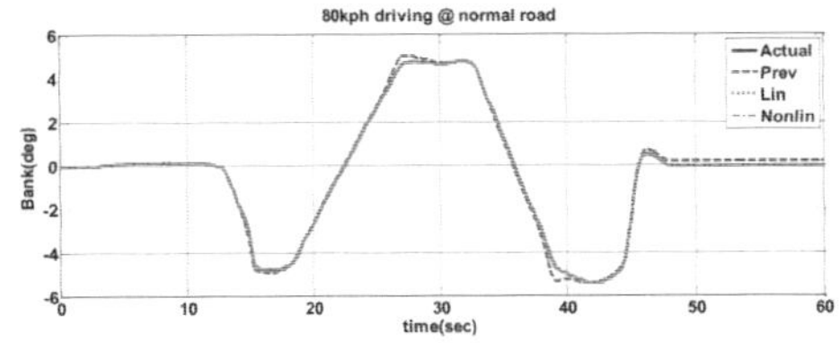

Figure 7. Case 2 Road bank angle

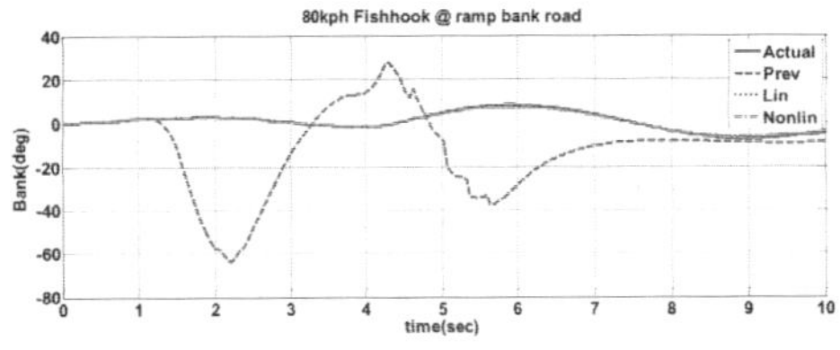

Figure 8. Case 3 Road bank angle

 https://doi.org/10.1051/matecconf/201816602008

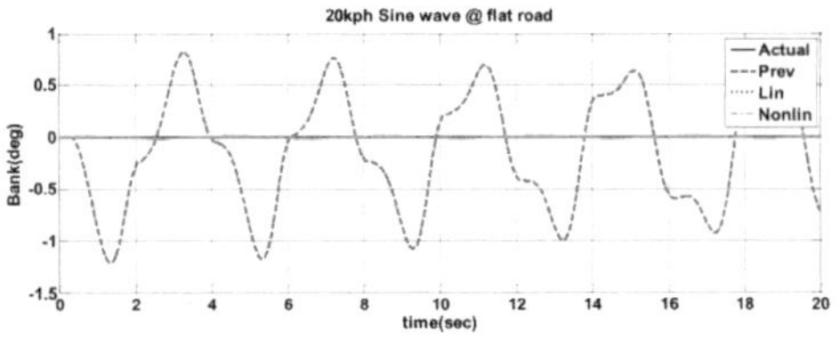

Figure 9. Case 4 Road bank angle

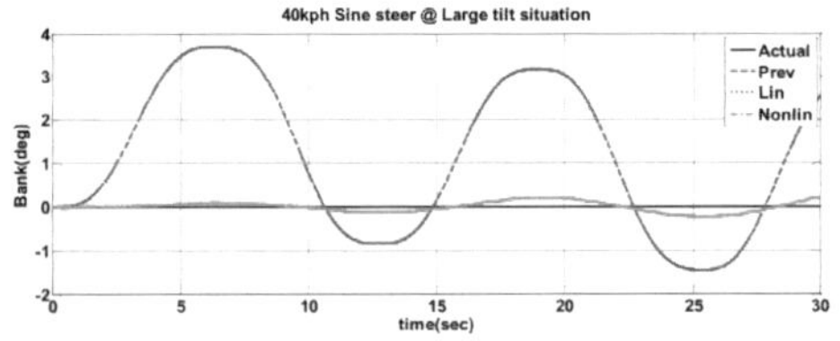

Figure 10. Case 5 Road bank angle

In this figures, "Actual" means simulation data. This data is considered actual vehicle test data. "Prev" means road bank angle estimation using only kinematic model based pseudo measurement such as those of previous result. "Lin" and "Nonlin" refer to the estimation result of linear and nonlinear estimators.

The figures above show the estimated road bank angle and the figures below show the estimated vehicle roll angle. From Case 1 to Case 3, there is few difference between road bank angle and vehicle roll angle. Therefore, the simulation results indicate that the vehicle suspension roll angle or vehicle tilt angle is small.

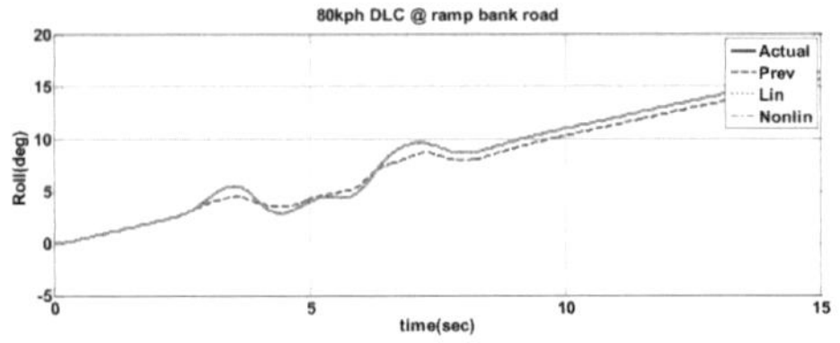

Figure 11. Case 1 Roll angle

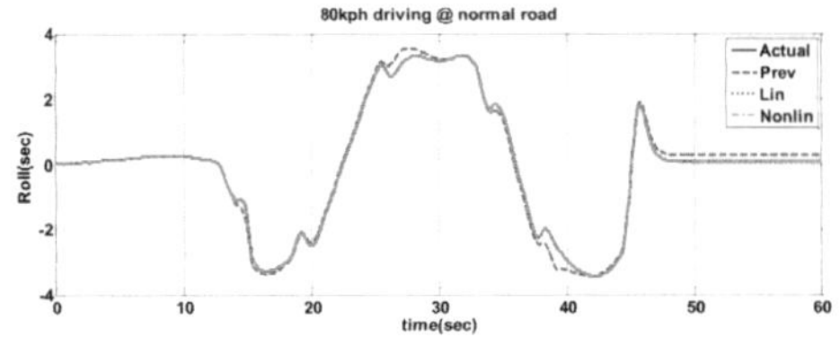

Figure 12. Case 2 Roll angle

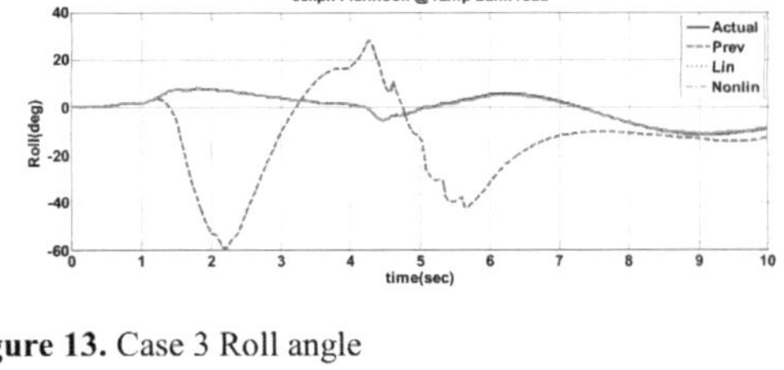

Figure 13. Case 3 Roll angle

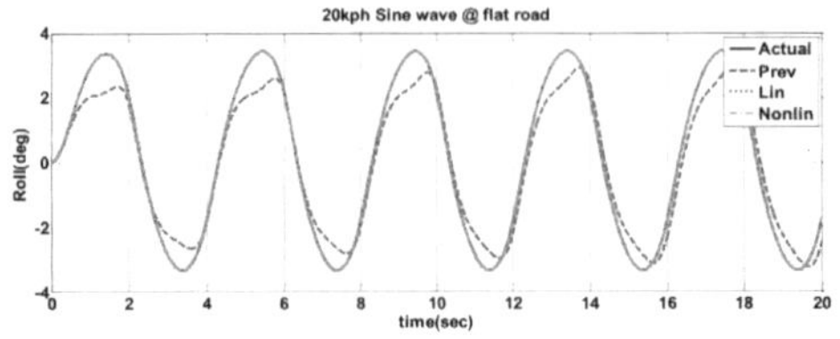

Figure 14. Case 4 Roll angle

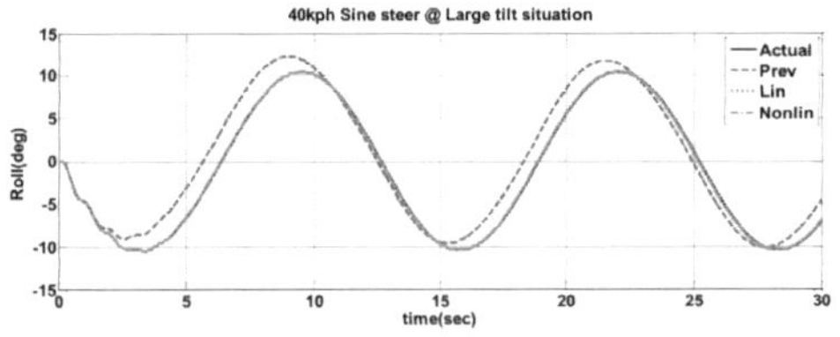

Figure 15. Case 5 Roll angle

Conversely, Case 4 and Case 5 roll and bank angle results show a significant difference between road bank angle and vehicle roll angle. Therefore, in this maneuver, the vehicle has a large inclination angle.

According to Case 2, in the normal driving situation, the proposed estimation method is slightly more accurate than the previous estimation. Since the tilt angle of the vehicle is small, the lateral dynamics are negligible. Therefore, the estimation results are almost unchanged. In the contrary, Case 1, Case 3, Case 4 and Case 5 show that the difference between the previous estimator result and proposed estimator results. Especially in case 4 and case 5, the performance of the previous estimator was significantly reduced. On the other hand, from Case 1 to Case 5, the proposed evaluator provides precise results for various maneuvers.

4 Conclusions

In this paper, the estimator which is specialized the three wheel tilting vehicle is proposed. Using vehicle dynamic and kinematic model, the accurate pseudo measurement can be obtained in both steady and transient states. Unlike previous study, lateral dynamics are not neglected, and more accurate estimation results can be obtained from extreme maneuver. Also, unlike previous studies, nonlinear estimators is proposed for accuracy with large tilt angle maneuver. Other prior vehicle roll estimators designed for conventional four wheel vehicles are not accurate when the vehicle's tilt angle is large. However, this estimator is more accurate because the proposed estimator in this paper considers the dynamics of a three-wheeled tilting vehicle. In conclusion, in this paper, the road angle estimator for three wheel tilting vehicle using 6D-IMU sensor is defined and verified. The main contribution of this paper is to propose a multi model estimator using steady state factor. This algorithm is suitable for three wheel tilting vehicle and verified using a custom ADAMS tilting vehicle.

Acknowledgement

This work was supported in parts by the National Research Foundation of Korea grant funded by the Korea government (MSIP) (No.2017R1A2B4004116), the Ministry of Science and ICT, Korea, under the Information

Technology Research Center support program (IITP-2017-2012-0-00628) supervised by the Institute for Information and communications Technology Promotion, and the BK21 Plus Program.

References

1. Hibbard, Robin, and Dean Karnopp. "Twenty first century transportation system solutions-A new type of small, relatively tall and narrow active tilting commuter vehicle." Vehicle system dynamics **25**.5 (1996):321-347.
2. Gohl, J., et al. "*Development of a novel tilt-controlled narrow commuter vehicle*." (2006).
3. Robertson, James W., Jos Darling, and Andrew R. Plummer. "Combined steering–direct tilt control for the enhancement of narrow tilting vehicle stability." Proceedings of the Institution of Mechanical Engineers, Part D: Journal of Automobile Engineering **228**.8 (2014): 847-862.
4. Piyabongkarn, D., T. Keviczky, and R. Rajamani. "Active direct tilt control for stability enhancement of a narrow commuter vehicle." International Journal of Automotive Technology **5**.2 (2004): 77-88.
5. Gohl, J., et al. "Active roll mode control implementation on a narrow tilting vehicle." Vehicle system dynamics **42**.5 (2004): 347-372.
6. Kidane, S., et al. "Road bank angle considerations in modeling and tilt stability controller design for narrow commuter vehicles." American Control Conference, 2006. IEEE, 2006.
7. Han, Kyoung Seok, et al. "Adaptive Scheme for the Real-Time Estimation of Tire-Road Friction Coefficient and Vehicle Velocity." IEEE/ASME Transactions on Mechatronics (2017).
8. Eric Tseng, H., Li Xu, and Davor Hrovat. "Estimation of land vehicle roll and pitch angles." Vehicle System Dynamics **45**.5 (2007): 433-443.

Design of Model-based Controller with Disturbance Estimation in Steer-by-wire System

Sanghun Jung[1], Seibum Choi[1,a]

[1,1a] *KAIST, Department of Mechanical Engineering, 34141 Daejeon, Republic of Korea*

Abstract. The steer-by-wire system is a next generation steering control technology that has been actively studied because it has many advantages such as fast response, space efficiency due to removal of redundant mechanical elements, and high connectivity with vehicle chassis control, such as active steering. Steer-by-wire system has disturbance composed of tire friction torque and self-aligning torque. These disturbances vary widely due to the weight or friction coefficient change. Therefore, disturbance compensation logic is strongly required to obtain desired performance. This paper proposes model-based controller with disturbance compensation to achieve the robust control performance. Targeted steer-by-wire system is identified through the experiment and system identification method. Moreover, model-based controller is designed using the identified plant model. Disturbance of targeted steer-by-wire is estimated using disturbance observer(DOB), and compensate the estimated disturbance into control input. Experiment of various scenarios are conducted to validate the robust performance of proposed model-based controller.

1 Introduction

X-by-wire (XBW) system is being introduced in the future automotive as the next generation automotive control system [1]. XBW system is being developed to provide the driver with the convenience of driving, steering, and braking. With this trend, Steer-by-wire (SBW) system is a future-oriented next-generation steering system with a large attention in the automotive industry. In SBW system, mechanical and hydraulic elements are removed. And electrical signal and actuator is used to control the steering system. Therefore, a driver and a wheel physically separated completely.

There are two main issues in SBW system control. First issue is applying reaction torque to the driver to improve the 'steering feel'. Due to the separation between driver and wheel, driver cannot feel the road information from wheel. Therefore, many studies are undergoing to apply reaction torque to compensate the 'steering feel'. Second issue is obtaining high control performance. Due to the separation, SBW system must have high tracking control performance.

In previous works, developing SBW control system is mainly focused on first issue ([2], [3], and [4]). Most of papers use simple PID controller as lower level controller to obtain the tracking performance. In the paper that proposes a controller that responds more strongly to disturbances, friction torque and self-aligning moment are modeled and added as feedforward term. However, this paper assumes that friction coefficient between road and wheel and vehicle body weight is constant. However, on vehicles with a light weight, such as micro-mobility, weight changes in wide range depending on the number of passengers. In addition, friction coefficient between road and wheel varies wide range with the road condition [5]. As a result, friction torque and self-aligning moment varies in wide range. Because of this variation of disturbance torques, constant feedforward term will not compensate enough for disturbance.

This paper proposes the robust lower level controller for SBW system. Targeted SBW system is modeled and identified through experiment results. Identified model is used to construct model-based controller. By using the model-based controller, we can achieve the fast response in transient input. Also, the friction torque and the self-aligning moment varying in a large range are treated as system disturbances. And we construct the disturbance observer (DOB) to estimate the system disturbances and compensate it to the control input. Thus, disturbance is rejected and we can achieve the robust control for various scenario. Proposed controller is validated through experiment using SBW system of 3-wheeler micro-mobility. Experiments are carried out according to the three road conditions that indicate friction torque varies. Results show that friction torque is well estimated, and performance of proposed controller is guaranteed regardless of road condition.

The rest of this paper is organized as follows. Section 2 presents the system modeling and identification method. Experimental setup, identification results are presented in this section. Section 3 presents the model-based controller with disturbance estimation design. Experimental result of proposed controller is presents in section 4. Conclusion is provided in section 5.

[a] Corresponding author: sbchoi@kaist.ac.kr

2 Steer-by-wire System Modeling & Identification

2.1 System description

In this paper, SBW system of light weigh 3-wheeler micro mobility is targeted. Targeted vehicle operated by 2 front in-wheel motors, and 1 rear steering motor.

To analyse and validate the proposed controller, we construct the test bench as Figure 1. Overall control algorithm is constructed using Matlab & Simulink. Constructed controller is built in to the MicroAutoBox to generate the control input and receive the measurement data.

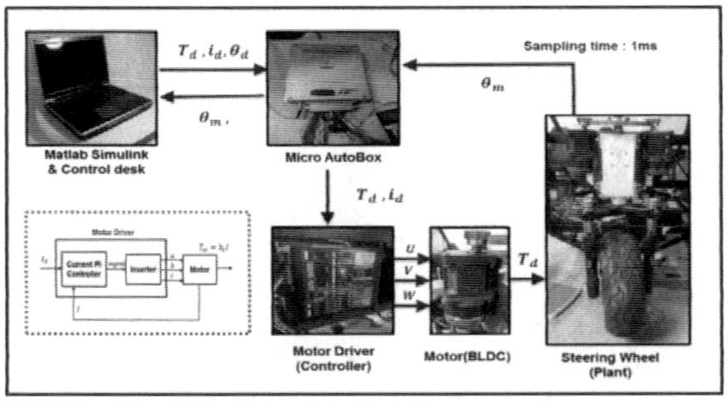

Figure 1. SBW system hardware and test bench set up

2.2 System modeling & identification

Mechanical system of SBW can modeled as simple 2nd order mass-damper system ([6], [7]). Two BLDC motors drive this system. Torque form motors is transferred through gearbox and rotate the steering wheel. So, we can summarize the system dynamics equation as following equivalent equation.

$$I_{eq}\ddot{\theta}_s + B_{eq}\dot{\theta}_s + T_{dist} = 2NT_m = u \tag{1}$$

Where

$$I_{eq} = I_s + 2N^2 I_m \tag{2}$$

$$B_{eq} = B_s + 2N^2 B_m \tag{3}$$

If we ignore the disturbance torque from ground, i.e. $T_{dist} = 0$, we can calculate the transfer function betweetn control input torque T and steering wheel angle θ as follow.

$$G(s) = \frac{\Theta(s)}{T(s)} = \frac{1}{I_{eq}s^2 + B_{eq}s} \tag{4}$$

In this research, we construct the model-based controller. Therefore, it is important to know the exact model information such as parameters or bandwidth. In the equation (4), system parameters are unknown. Therefore, we use the closed-loop system identification method to estimate the system parameters. The closed loop system transfer function is calculated as follow.

$$\frac{\Theta(s)}{\Theta_d(s)} = \frac{KG(S)}{1+KG(S)} = \frac{K}{I_{eq}s^2 + B_{eq}s + K} \tag{5}$$

Where θ_d is commanded steering angle, θ is actual steering angle, and K is the feedback gain of closed loop system. To analyze the frequency response of the system, we apply 0 to 5 Hz sine wave as command input. Figure 2 show the command input and following results.

Through frequency response of the system, we use the empirical transfer function estimate(ETFE) to estimate the transfer function. Using the discrete fourier transform (DFT), we compare the ratio phase between commanded input and acutal ouput angle. As a results, we can calculate the bode plot form experiment data and use it to the system identification. Figure 3 shows the result of ETFE. We apply the command input in 0 to 5 Hz, so we do not use the information over 5Hz range of calculated bode plot. Finally, we can obtain system parameters I_{eq} and B_{eq}.

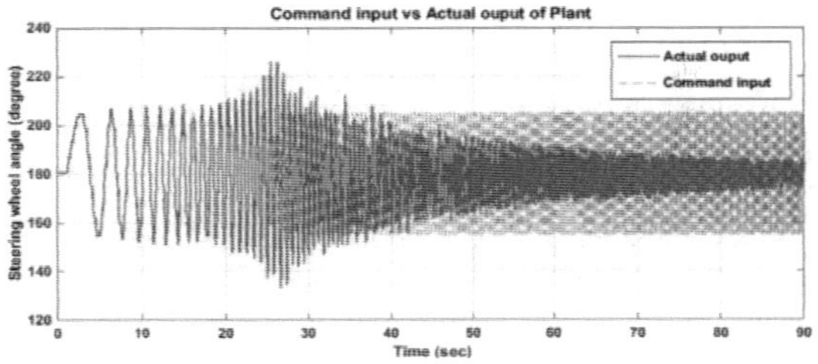

Figure 2. Experiment result: Command and actual angle for system identification.

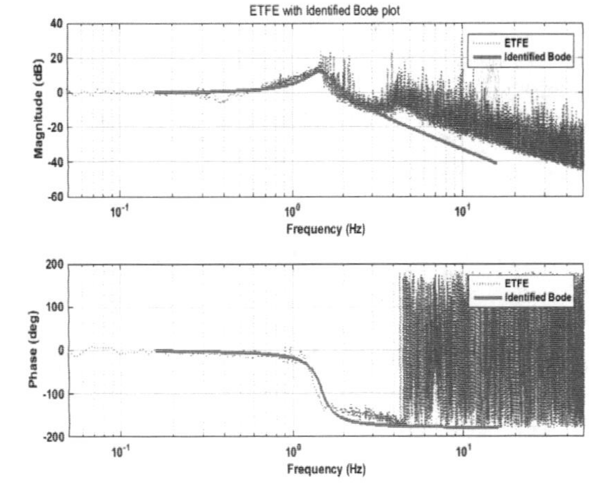

Figure 3. ETFE with identified bode plot

3 Model-based controller design with disturbance estimation

3.1 Model-based controller design

In previous section, targeted SBW system is modeled in 2nd order system and well identified. Therefore, we can construct the model-based controller using identified model.

Control target of the system is track the real steering wheel angle θ to desired command angle θ_d. Thus, we can set the error as following.

$$e = \theta_d - \theta \qquad (6)$$

Fianl control target is to make the error equation converge to zero as time passed, error dynamic is constructed as folloing equation.

$$\ddot{e} + 3\lambda\dot{e} + 3\lambda^2 e + \lambda^3 \int e = 0 \qquad (7)$$

The order of dynamic equation of the system is 2 as equation (1) shows. Generally, the order of error dynamics follows the system order. But, in this research, targeted system has a feature of wide range of large friction. Generally, friction in 2nd order system causes the steady-state error. So, we apply the integral term to the 2nd order error dynamic so that we can handle the steady-state error from friction. Furthermore, we place the pole of gains in single point, we can tune the gain more conveniently.

Combining system dynamic equation (1) and error dynamics (7), we can organize the equations as follow.

$$\ddot{e} = \ddot{\theta}_d - \ddot{\theta} = \ddot{\theta}_d - \frac{1}{I_{eq}}\left(u - B_{eq}\dot{\theta} - T_{dist}\right) = -3\lambda\dot{e} - \beta\lambda^2 e - \lambda^3 \int e \qquad (8)$$

The above equation can be summarized in terms of u as follow.

$$u = I_{eq}\ddot{\theta}_d + B_{eq}\dot{\theta} + T_{dist} + 3\lambda^2 I_{eq} e + 3\lambda I_{eq}\dot{e} + \lambda^3 I_{eq}\int e \qquad (9)$$

In above equation, first part of input is feed-forward control based on identified model. And T_{dist} is unknown disturbance from ground, and rest term is formed of PID feedback controller.

Model-based feed-forward controller calculate the control input from desired command signal immediately, it could obtain response that is much more fast rather than feedback control only. If the model is not accurate, system can be unstable. However, as shown in previous chapter, we assume that system is well modeled and well identified. Therefore, we proposed the model-based controller.

3.2 Disturbance estimation using disturbance observer

In previous section, model-based controller is constructed. However, the disturbance T_{dist} is still unknown factor. As mentioned before, disturbance of SBW system is dominant factor of control. Thus, disturbance compensation logic is strongly requried. In this paper, we use the disturbance observer(DOB) to estimate and compensate the unknown disturbances.

The simple DOB structure of single input – single output(SISO) system is shown in Figure 4[8]. The notation P, P_n indicate transfer functino of real plant and nominal plant respectively.

The final output y can represented as function of reference input r and disturbance d as follow.

$$Y(s) = T_{yr}(S)Y_r(s) + T_{yd}(s)D(s) \qquad (10)$$

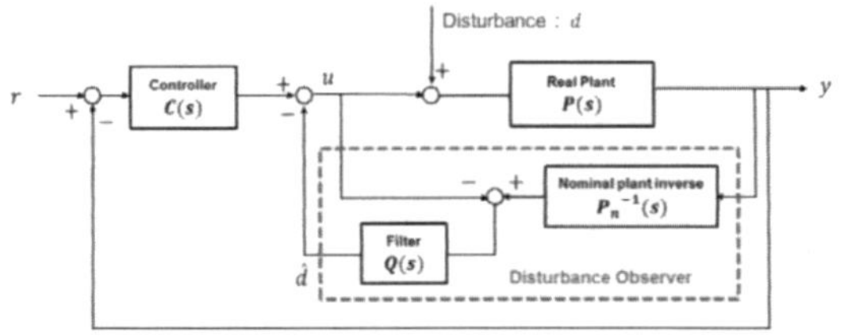

Figure 4. Basic structure and block diagram of DOB

T_{ry}, T_{dy} is closed loop transfer function betweeen r, d and output y respectively. We can conclude that if Q-filter $Q(s)$ is well designed, we can acheive folloing results for low frequency range input under cutoff frequency ω_q

$$T_{ry}(s) = \frac{P_n(s)C(s)}{1 + P_n(s)C(s)}, \; T_{dy}(s) = 0 \qquad (11)$$

This result means that SISO system with disturbance behaves same as nominal plant SISO system without disturbance. So disturbance is rejected internally by disturbance observer.

Applying above result in targeted SBW system, we can construct following euqations.

$$I_{eq}\ddot{\theta} + B_{eq}\dot{\theta} + T_{dist} = u \qquad (12)$$

$$I_n\ddot{\theta} + B_n\dot{\theta} = u + \hat{d} \qquad (13)$$

The estimated disturbancd $\hat{d}$ is solved by above simultaneous equations. The result is as follow.

$$\hat{d} = -\Delta I\ddot{\theta} - \Delta B\dot{\theta} - T_{dist} \qquad (14)$$

Where ΔI and ΔB represent the difference between real and nominal plant. If we assume that system is well identified so that $\Delta I = \Delta B = 0$, estimated disturbance $\hat{d}$ converges to $-T_{dist}$. Thus, if we combine this result and constructed model-based controller in (9), final form of model-based controller is made as follow.

$$u = I_{eq}\ddot{\theta}_d + B_{eq}\dot{\theta} - \hat{d} + 3\lambda^2 I_{eq} e + 3\lambda I_{eq}\dot{e} + \lambda^3 I_{eq}\int e \qquad (15)$$

Therefore, controller composed of model-based feed forward controller, disturbance compensations term, and feedback controller.

4 Experimental results

3.1 Model-based controller design

We conduct the experiment to validate the proposed model-based controller with disturbance estimation. To validate the robust performance of controller, we construct the three scenarios that represent the various friction torque-applying situation. The disturbances of SBW system composed of friction torque and self-aligning moment as mentioned before. However, the experiment only deals with stationary vehicle scenarios, which means that the disturbance is composed of friction uniquely. Friction torque between wheel and road can be analyzed as follow.

$$T_{fric} = k \cdot \mu F_z \quad (16)$$

The notation k , μ , and F_z represent the constant, friction coefficient, and vehicle vertical load respectively. Equation (20) implies that friction torque can varies by changing the firciton toruqe coefficient or vehicle vertical load. Thus, we construct the three scenario with three different vehicle vertical loads. Because vertical load is much more easier to measure and analyze rather than friction coefficient. Table 1 shows the scenario and situation that each scenario implies.

The case 1 refers to the normal driving situation. The case 2 refers to the system with higher disturbance in the case of high-mu surface or vehicle with heavy load situation. The case 3 refers to the system with lower disturbance in the case of low-mu surfaces, such as ice or snow road. The case 3 was carried out by lifting the steering wheel.

Table 1. Experiment scenarios and implying situation.

No.	Measured F_z	Case information
1	100 kgf	Normal driving
2	145 kgf	High-mu surface, loaded vehicle
3	0 kgf	Low-mu surface: ice, snow road

It can be said that propsed controller has robust control performance if the controller performance guranteed through the above scenarios with single gain and without additional gain-scheduling.

The experiments are carried out acoording to the each scenario above applying the sinusoidal wave with 0.3Hz and amplitude 30 degree. To confirm the effect of model-based feedforward control and disturbance compensation logic, we conduct the experiment for each case with two differenct controller ;(a) only feedback control, (b) model-based foeedforward control with disturbance estimation. Control input of each case is shown as follow.

(a) $$u = 3\lambda^2 I_{eq} e + 3\lambda I_{eq} \dot{e} + \lambda^3 I_{eq} \int e \quad (17)$$

(b) $$u = I_{eq} \ddot{\theta}_d + B_{eq} \dot{\theta} - \hat{d} + 3\lambda^2 I_{eq} e + 3\lambda I_{eq} \dot{e} + \lambda^3 I_{eq} \int e \quad (18)$$

During the experiment, feedback gains are fixed. The experiment reuslts is shown as follows.

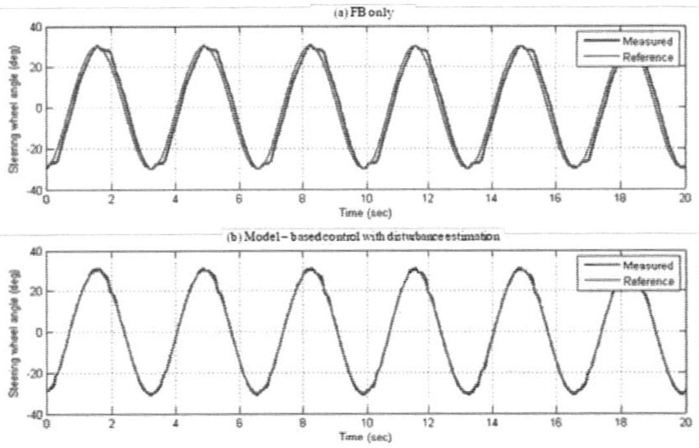

Figure 5. Experiment results: Scenario 1. Normal driving situation. (a) FB only (b) model-based control with disturbance estimation.

Experiment result for scenario 1 show the effect of model-based controller. Only feedback control, as shown in (a), output signal is lagged and suppressed due to the friction. Thus, we can show that reaction rate is lowered. However, as shown in (b), compensating the model-based feedforward term and disturbance estimation results make reaction rate much faster than (a). Therefore, tracking performance is improved.

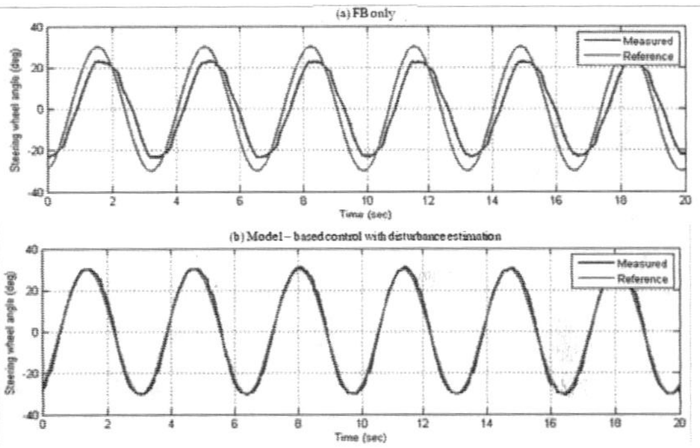

Figure 6. Experiment results: Scenario 2. High-mu surface or loaded vehicle. (a) FB only (b) model-based control with disturbance estimation.

The performance of the proposed controller is more prominent in scenario 2. This scenario implies larger friction, graph (a) shows the poor tracking performance. Due to the large friction, system output suppressed a lot and response rate becomes low. However, a term that compensates for the disturbance is included in the proposed controller. Thererfore, control ouput of proposed controller shows good perfermance with same feedback gains.

https://doi.org/10.1051/matecconf/201816602009

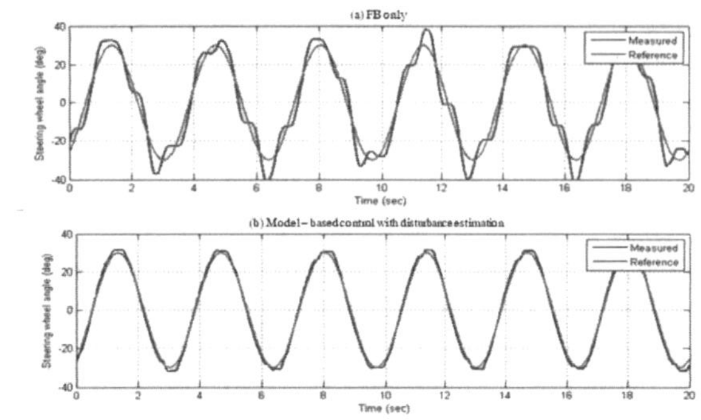

Figure 7. Experiment results: Scenario 3. Low-mu surface. (a) FB only (b) model-based control with disturbance estimation.

Scenario 3 indicates the frictionless system, which means a system without disturbances. Friction torque is the major part of the system as mentioned before. Thus, feedback gain is adjusted high-gain. However, if the friction disappears in that state, the systems becomes unstable as shown in (a). As mentioned in section 3.2, disturbance observer makes the system to behave as nominal plant. Therefore, if the gain is adjusted appropriately so that the nominal plant can operate in stable region, the proposed controller can obtain good tracking result. The result shown in (b).

As a results, propsed controller can obtain good tracking performance in various scenarios with fixed feedback gain. Therefore, we can conclude that proposed controller has robust control performance in SBW system. Table 2 shows the RMS error for each case of expriement.

Table 2. RMS error of proposed controller

No.	Measured F_z	RMS error (deg)
1	100 kgf	0.6924
2	145 kgf	1.2338
3	0 kgf	1.3248

5 Conclusion

In light weight micro-mobility, disturbance is major part of the SBW system control. Therefore, disturbance compensation logic was strongly demanded. In this paper, we proposed the model-based controller with disturbance estimation in SBW system. Experiments are conducted to validate the proposed controller. Three scenarios representing the various disturbance are selected to validate the robustness of the proposed controller. As a result, model-based controller with disturbance estimation obtain robust control performance in various scenario with fixed feedback gain. The results of this research are expected to improve the performance of steering control of future-Micro-Mobility vehicle.

Acknowledgment

This work was supported in parts by the National Research Foundation of Korea grant funded by the Korea government (MSIP) (No.2017R1A2B4004116), the Ministry of Science and ICT, Korea, under the Information Technology Research Center support program (IITP-2017-2012-0-00628) supervised by the Institute for Information and communications Technology Promotion, and the BK21 Plus Program.

References

1. Y.H.Ki, I.Oh, J.Jeon, J.Chun, H.S.Ahn, *A new type of* clamping *force estimator for electromechanical brake systems,* IEEE Transactions on Mechatronics, Vol.**10**, December, (2005)
2. Yung-Hsiang Judy Hsu, *Estimation and Control of Lateral Tire Forces Using Steering Torque*, Ph.D Dissertation, Stanford University, U.S, (2009)
3. Seok-Hwan Jang, Tong-Jin Parkl chang-soo Han, *A control of vehicle using steer-by-wire system with hardware-in-the-loop-simulation system,* IEEE Transactions on Mechatronics, Vol.**10**, December, (2005)
4. Hiraku Ogura, Toshiyuki Murakami, *Reaction Force Control on Accelerator Pedal and Steering Wheel based on Tire-Road Friction Forces*, 35th Annnual Conference of IEEE industrial Electronics, (2009)
5. Hiraku Ogura, Toshiyuki Murakami, *Reaction Force* Control *on Accelerator Pedal and Steering Wheel based on Tire-Road Friction Forces*, 35th Annnual Conference of IEEE industrial Electronics, (2009)
6. Sanket Amberkar, Farhad Bolourcho, *A Control System Methodology for Steer by Wire Systems,* SAE technical paper series, Steering and Suspension Technology Symposium (2004)
7. Paul Yih and J.Christian Gerdes, *Modification of vehicle handling characteristics via Steer-by-Wire,* IEEE Transactions on Control System Technology, Vol.**13**, No.6, November, (2005)
8. Shihua Li, Jun Yang, Wen-Hua chen, Xisong chen, "Disturbance Observer-Based Control, Methods and Applications", *CRC Press*, (2014)

Numerical Study of Jet Impingement Heat Transfer on A Roughened Flat Plate

Alenezi A.H.[1], Abdulrahman Almutairi [2], Hamad M. Alhajeri[3] and Addali A [4]

[1]*Mechanical Power and Refrigeration Department, College of Technological Studies, Shuwaikh, Kuwait*
[2]*Mechanical Power and Refrigeration Department, College of Technological Studies, Shuwaikh, Kuwait*

Abstract. The purpose of this research paper is to investigate the effect of surface roughness on heat transfer. This was achieved by investigating the heat transfer in two cases: a smooth, horizontal surface (the baseline case) and the same surface with a roughness element added to it. The roughness elements took the shape of cubical pin-fin. The roughness element was further investigated by varying its height and width (e by e) to study their impact on the average Nu. Results are presented in the form of average Nusselt number $\overline{N_u}$ within and beyond the stagnation region. Each roughness element was arranged in a circle concentric with the geometric centre (i.e. centre of jet) with a radius of one and a half jet diameters (R/D = 1.5) The jet diameter kept constant for all simulations (D=13.5mm) where the plane was located at H/D = 6 below the jet, for all the tests with a jet Reynolds number of 20,000 and jet temperature is 20°. The cubical pin-fin was tested for six different heights (e) from 0.25 mm to 1.50 mm in incremental steps of 0.25 mm for jet angle (α) of 90°.

1 Nomenclature

D	=	Jet diameter (m)
R	=	radial distance (m)
e	=	element height and width (m)
$\boldsymbol{\alpha}$	=	jet angle (°)
Re	=	Reynold's number
Nu	=	local Nusselt number
r/D	=	dimensionless radial distance
$\overline{N_u}$	=	average Nusselt number

2 Introduction

Jet Impingement cooling is a complex technique introduced into GT blade cooling in the early 1960's and has proven to be very effective for increasing the heat transfer rate compared to other cooling techniques. It is applied mostly on the inner surface of the blade through small holes in the inner passages to directly impinge on the hot regions such as blade leading edge Kanokjaruvijit, and Martinez-botas [1],[2] added dimples to the surface of the target plate to examine the effect of jet impingement on a surface with a staggered array of dimples. They tested two dimpled geometries: cusped elliptical shape and hemispherical shape. The combined coupling effect of jet impingement and channel flow with the presence of cross-flow provided the highest heat transfer performance of the dimpled plate at large jet to plate spacing. At moderate cross-flow, either the channel flow or jet impingement could dominate the heat transfer rate. At low cross-flow, jet impingement was found to be the dominant heat transfer mechanism. With the presence of dimples, flow circulation inside the concave dimples reduced heat transfer, especially with the smaller jet to plate spacing.

Zhang, et. al. [3] in his experimental and numerical studies employed an orthogonal single jet on a protrusion heated surface (constant heat flux). The protrusion took a shape of spherical pin-fin with height (δ) and diameter (D). Results were reported for local Nu distributions for Re 5000, 10,000 and 23000 and for normalized protrusion heights (δ/D) of 0.1,0.2 and 0.3. In general, flow structures are not effected by both protrusion depth or jet Reynolds number. A diversion in flow direction occurs when the flow meets with the protrusion and low pressure wake area behind the rib occurs where its length depends on the protrusion height. Overall, distributions of local Nu show higher magnitude for roughened surface when compared to smooth surface.

Kim, and Lee [4] conducted an experimental study of heat transfer and fluid flow characteristics of water drop impingement on a heated porous surface. They tested four different porous surfaces based on permeability and surface roughness obtained by using different sizes of small glass beads. Their test demonstrated two primary regions of impingement: contact and non-contact regions. The contact region is considered as water drop evaporation or boiling when in contact with the surface. The non-contact region is a surface film boiling regime where the drop was suspended in the air. The effects of wall temperature, impact velocity and particle size of the porous substrate on thermal and fluid flow performances were investigated.

Celik [5] compared a smoothed and roughened the heated surface with co-axial jet impingement. The co-axial jet was a double pipe arrangement. This study concluded that the co-axial jet gave higher heat transfer than a single circular jet and that rough surface promoted better heat transfer due to interruption of the boundary layer over the roughened plate. The roughened surface achieved up to 27% higher averaged Nusselt number using the co-axial jet compared to a single circular jet configuration.

3 Numerical setup

3.1 Computational domain and boundary conditions

Figure 1 shows the axisymmetric computational domain and boundary conditions used in this simulation. The domain length and Radius which include the heated impingement surface (also referred to in the literature as "wall") is 20D. An axisymmetric assumption is made to save computational time and cost; thus, we need to consider only one half of the domain shown in the figure. A separate simulation was carried out to provide the velocity profile on the domain inlet to ensure a fully developed flow based on the Reynolds number used in this simulation.

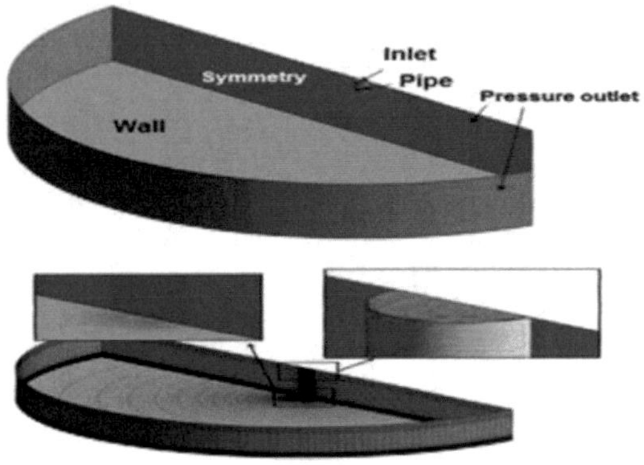

Figure 1. Geometry, boundary conditions, and grid for an orthogonal jet impingement.

3.2 Mesh generation, turbulence modelling and fluid properties

Figure 2 below shows an array of cubical pin-fins with detailed mesh for one of them. The array has also radius R of 1.5D measured from the geometric centre. The same as the case of spherical pin fin, a fixed distance of 1.3mm kept between all elements with varying their number to a maximum of 50 roughness elements when e=0.25mm, all have same width and height of e. The maximum percent increase in total surface area by using cubical pin-fin is 0.5%

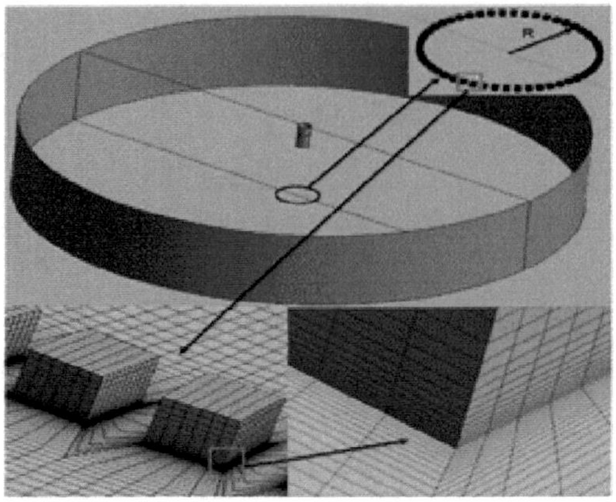

Figure 2. Computational mesh for a square cross-sectional pin-fin roughness element on a plane wall

The detailed geometry of all cubical pin-fins is shown in table 1.

Table 1. Detailed geometry for cubical pin-fin

Height e (mm)	Area (mm^2) (5 faces)	Pitch (mm)	No. of roughness elements
0.25	0.3125	2.3	30
0.5	1.25	2.05	30
0.75	2.8125	1.79	30
1	5	1.55	30
1.25	7.8125	1.3	30
1.5	11.25	1.05	30

4 Results and discussion

4.1 Grid refinement study

This section shows the validation of the 3D- the axisymmetric model used in this simulation against the benchmark experimental result of O'Donovan and Murray [6].

Generally, both numerical solution accuracy and computational time depend primarily on mesh refinement. A suitable grid is one which has an acceptable run-time and good accuracy. A grid analysis was used in this research paper to certify that the solution is independent of the computational grid where three grid cases were used in the grid refinement study as shown in table 2 below.

Table 2. Detailed geometry for cubical pin-fin

№	Size, cells	Y+
Mesh 1	400 000	0.5
Mesh 2	986 000	0.5
Mesh 3	1 783 000	0.5

The commercial tool Ansys Fluent 16.2 was employed in this study to simulate the jet impingement. Extra attention was taken on the near-wall region since it plays an important role for convective heat transfer. The SIMPLEC scheme and Green-Gauss Cell Based gradient for spatial discretization were employed in this study using second order discretization schemes energy and momentum equations to produce more accurate results for heat transfer where first order schemes were used for other equations. Several steps were used before finalizing the choice of the appropriate solution: first, the use of the entire domain initialized by the inlet flow conditions employing first order upwind discretization to reach a convergence criterion at 10^{-6} for energy equation and 10^{-4} for the rest. The second step is seeking the solution by mixing up different orders of discretization schemes.

Figure 3 shows the simulated local Nusselt number varied with the radial distance (r/D) using the RNG k-ε turbulence model, for different grids and α=90°. The numerical results were compared against the experimental benchmark work of [6] . It is notable, that both experimental and the numerical local Nusselt number values are close to all grid cases and therefore, mesh 2 was adopted for the remaining numerical calculations.

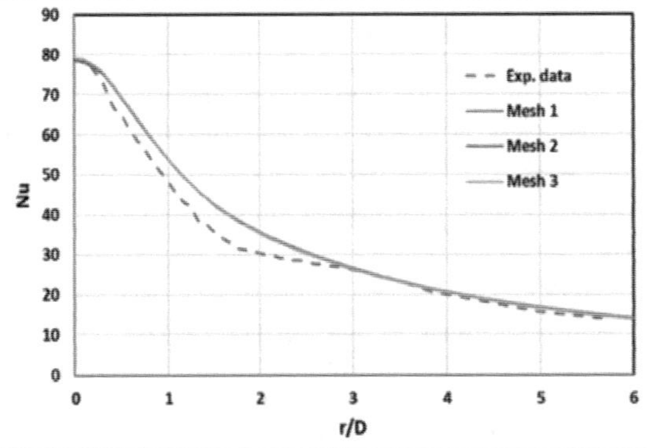

Figure 3. Local Nusselt number distribution for three grids, H/D=6, Re=10,000, α=90°

4.2 Average heat transfer performance

Figure 4 shows the effect of cubical height on the local Nusselt number distribution for a circular jet impinging on a flat plate. The cubical roughness element has a length and a width of e. Overall, two peaks of local Nu are shown for all heights due the flow recirculation occurs in front and after the roughness elements. Both peaks increase as the height increases so the drop in local Nu which occurs in the region attached to the roughness wall. At r/D≈2.25, and e≤0.75mm, all local Nu curves starts to match as the height variation is no more affects the heat transfer rate. However, for e≥1mm, curves of roughened surfaces show a slight lower value of local Nu than the smooth case.

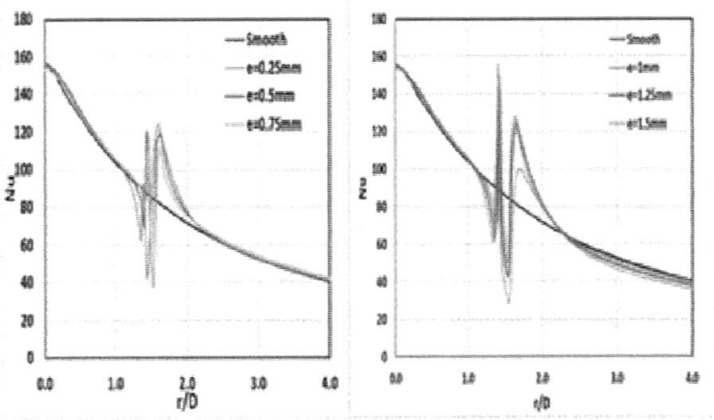

Figure 4. Effect of cubical pin-fin height on local Nusselt number for Re= 20,000, H/D=6 and α = 90°.

Local Nu contours are shown in Figure 5 below. Overall, cubical pin-fins show higher local heat transfer distributions for all heights. As e increases, so does the local heat transfer rate, this is obvious for 0.25mm≤e≤1.25mm. However, for e=1.5mm, Nu starts to fall due to the increase of the wake region behind the roughness element. This low-pressure region forces the flow to recirculate within it which effects the main flow velocity by reducing it. This reduction causes heat transfer rate to diminish.

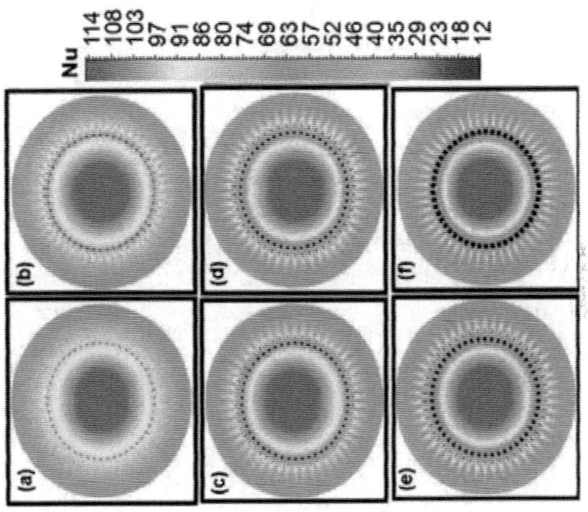

Figure 5. Local Nu contours for roughness heights of (a) 0.25 mm, (b) 0.5 mm, (c) 0.75 mm, (d) 1mm, (e) 1.25mm and (f) 1.5 mm, Re= 20,000, H/D=6 and α = 90°.

To identify the low-pressure zone behind the roughness element, contours of velocity vectors of three cubical pin fins heights are shown in figure 6 below. The figure shows velocity vector contours for three pin heights; 0.25 mm, 0.75 mm and 1.5 mm. As shown in the figure, the higher the pin, the larger the separation region (the region above the fin top edge), the larger the low-pressure region, and the higher the local Nu. However, for e=1.5 mm, flow starts to recirculate above the top edge of fin, as shown in the figure, causing the main jet to travel a longer distance and reducing its arrival velocity before it hits the heated surface resulting lowest increase in $\overline{N_u}$.

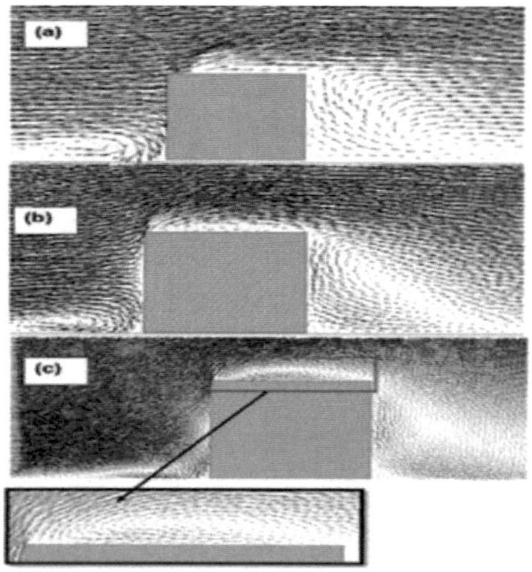

Figure 6. Contours of velocity vectors for heights (a) e=0.25mm, (b) e=0.75mm and (c) e=1.5mm, Re=20,000, H/D=6 and $\alpha = 90^{o}$

Figure 7 below shows the effect of using a cubical pin-fin on $\overline{N_u}$. The figure shows that, $\overline{N_u}$ can vary by up to about 7.2% of the value of the smooth surface, for the smaller pin heights (0.25 mm and 0.5 mm) but gradually decreasing to about -1.4% for e=1.5 mm. The rate of heat transfer increases as the tip of the square pin-fin gets closer to the jet exit since the surface area is increased. However, as the fin height further increases, both velocity and pressure of the jet decreases causing the average Nu over the surface area to diminish but the local Nu which is based only on the local value may increase [7].

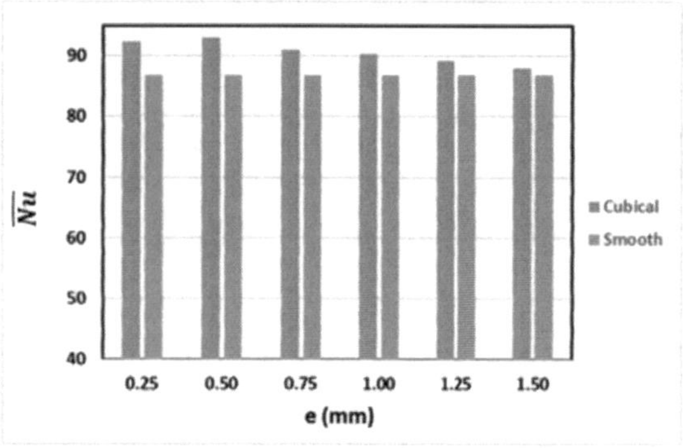

Figure 7. Effect of cubical pin-fin height on average Nusselt number over 0≤r/D≤4, Re= 20,000, H/D=6

5 Conclusion

The research paper reports the results of using simulation to investigate the effects on the average predicted Nu, of cubical pin-fins imposed on a flat, horizontal plane (the wall) subject to flow of cold fluid from a nozzle of diameter 13.5mm. The roughness element was located at the same radial distance measured from the geometric centre, R/D=1.5 on a plane located H/D=6. The optimum height of the spherical pin-fins was investigated using six heights; 0.25 mm to 1.50 mm in increments of 0.25 mm, for jet angle (α=90°).

The maximum enhancement of $\overline{N_u}$ was obtained by the use cubical pin-fin roughness elements with height (e) of 0.5 mm with 7.2% percentage increase comparing to the smooth case. The minimum enhancement percentage of the average Nu was 1.4% when employing cubical pin-fins with the maximum height (e) of 1.5mm. The enhancement percentages vary with changing the roughness element height and width.

6 References

1. Kanokjaruvijit, K., Martinez-Botas, R. F. Jet impingement on a dimpled surface with different crossflow schemes. International Journal of Heat and Mass Transfer. 2005; 48(1): 161–170.
2. Kanokjaruvijit, K., Martinez-Botas, R. F. Heat Transfer and Pressure Investigation of Dimple Impingement. ASME Conference Proceedings. 2005; 2005: 717–728.
3. Zhang D., Qu H., Lan J., Chen J., Xie Y. Flow and heat transfer characteristics of single jet impinging on protrusioned surface. International Journal of Heat and Mass Transfer. Elsevier Ltd; 2013; 58(1–2): 18–28.
4. Kim, W., S., Lee, S. Y. Behavior of a water drop impinging on heated porous surfaces. Experimental Thermal and Fluid Science. Elsevier Inc.; 2014; 55: 62–70.
5. Celik, N. Effects of the surface roughness on heat transfer of perpendicularly impinging co-axial jet. Heat and Mass Transfer/Waerme- und Stoffuebertragung. 2011; 47(10): 1209–1217.
6. O'Donovan, T., S., Murray, B. D. Jet impingement heat transfer - Part I: Mean and root-mean-square heat transfer and velocity distributions. International Journal of Heat and Mass Transfer. 2007; 50(17–18): 3291–3301.
7. Sanyal, A., Srinivasan, K., Dutta P. Numerical study of heat transfer from pin-fin heat sink using steady and pulsated impinging jets. IEEE Transactions on Components and Packaging Technologies. 2009; 32(4): 859–867.

Research and Design of Fixed Photovoltaic Support Structure Based on SAP2000

Xingxing Wang[1, 2], Guangjian Ji[1, 3], Hai Gu[2], Shuaishuai Lv[1, 2], Hongjun Ni[1, 2], Ping Wang[3], Ke Chen[1], Yue Meng[1]

[1] *School of Mechanical Engineering, Nantong University, Nantong, Jiangsu, 226019, P.R. China*
[2] *Jiangsu Key Laboratory of 3D Printing Equipment and Application Technology, Nantong Institute of Technology, Nantong, Jiangsu 226002, P.R. China*
[3] *Engineering R & D Center, Jiangsu Lead Aluminum Co., Ltd, Haian, Jiangsu, 226600, P.R. China*

Abstract. In the solar photovoltaic power station project, PV support is one of the main structures, and fixed photovoltaic PV support is one of the most commonly used stents. For the the actual demand in a Japanese photovoltaic power, SAP2000 finite element analysis software is used in this paper, based on Japanese Industrial Standard (JIS C 8955-2011), describing the system of fixed photovoltaic support structure design and calculation method and process. The results show that: (1) according to the general requirements of 4 rows and 5 columns fixed photovoltaic support, the typical permanent load of the PV support is 4679.4 *N*, the wind load being 1.05 kN/m^2, the snow load being 0.89 kN/m^2 and the seismic load is 5877.51 *N*; (2) by theoretical calculation of the two ends extended beam model, the beam span under the rail is determined 2200 *mm*; (3) by the way of using the single factor experiment, through the calculation and analysis of SAP2000, the three best supporting points of the support of the W stent are determined; (4) by comprehensive simulation, the optimal parameters for the rail, beam, support and bolt are 60× 60× 1.0, 60× 60× 1.0, 40× 50× 2.0, and M10 respectively.

1 Introduction

Solar energy as one of the new energy resources is most widely used currently. In recent years, with the popularization and application of solar photovoltaic support system, the design and development of structure of the photovoltaic support system have also become the focus of attention. At present, the photovoltaic support is mostly steel structure in the market, but the aluminum profile has the characteristics of light weight, beautiful appearance, corrosion resistance and other characteristics, which has attracted the attention of the market [1-4]. Compared with the automatic tracking support, the fixed photovoltaic support has smaller footprint, lower initial investment and less maintenance in the later stage of the support system; In structure, under the same environmental conditions, the stability of the fixed support structure is better than that of the tracking support; In the power generation capacity, the power generation of the tracking support is better than the fixed support, but it needs a longer investment recovery cycle[5-7]. The software SAP2000 has strong functions, such as model (2D, 3D model, etc.), editing function (adding and deleting units, copying and deleting, etc.), analysis function (time history analysis, dynamic analysis), load function (node load, rod load, surface load and temperature load), custom features, and design function etc.

Taking an engineering project in Japan as an example, the SAP2000 software was used in this paper to carry out the analysis and research on the bearing capacity of the fixed photovoltaic support under various load conditions, so as to provide a reference method for the structural design of the fixed photovoltaic support.

2 Simulation

2.1 The basic parameters of project

A project was located in Kaseda City Jinfeng town of Japan. The degree of the design angle of PV modules was 15, and the PV module specification was 1650*mm* ×991*mm*×40*mm*. The single photovoltaic array unit was composed of 20 photovoltaic modules, which were arranged into 4 rows and 5 columns. According to the design requirements of power station, in the photovoltaic support design process, the array structure strength should meet the environmental requirements, such as the wind load 1.05 kN/m^2, the snow load 0.89 kN/m^2, and the basic parameters were shown in table 1.

2.2 Design of overall scheme

(1) Design of photovoltaic support structure

According to the 4 rows and 5 columns PV modules of the fixed photovoltaic support overall requirements, combined with the project development experience, the triple-layer composite of photovoltaic support were rail, beam, and column; The conventional screw pile was used in the foundation part; At the same time, the rail and beam length were determined which were 8390 *mm* and 4300 *mm*; There were 5 rails arranged on the upper and lower ends of the PV module; The beam was vertically arranged at the underside of the rail, and the number and span of the beam needed to design and optimize; The type of the pillar arranged below the beam was W, and the 3 contact points of the beam needed to design and optimize.

Table 1. Parameters of PV module and design requirements of PV support

Parameter type	Parameter values
Module size	1650 *mm*×991 *mm*×40 *mm*
Module weight	19 *kg*
Module surface area	1.63515m^2
Mounting angle of PV support α	15°
Module height from the ground	1000 *mm*

(2) Lightweight design of photovoltaic stent

The commonly used sections of rail, beam, and column were 60×60, 60×60, and 40×50, and various thickness specifications could be selected. The photovoltaic support section could be seen in figure 1. The overall scheme of photovoltaic support structure and the type of section of the main profile were determined, and reducing the amount of aluminum material of the photovoltaic support was the main goal of lightweight design, under the premise of ensuring the structural strength of the photovoltaic support. Using the method of layer by layer design and verification discussed the effects on bearing capacity of the rail and beam with different thickness sections, and provided a basis for the selection of reasonable economic profile.

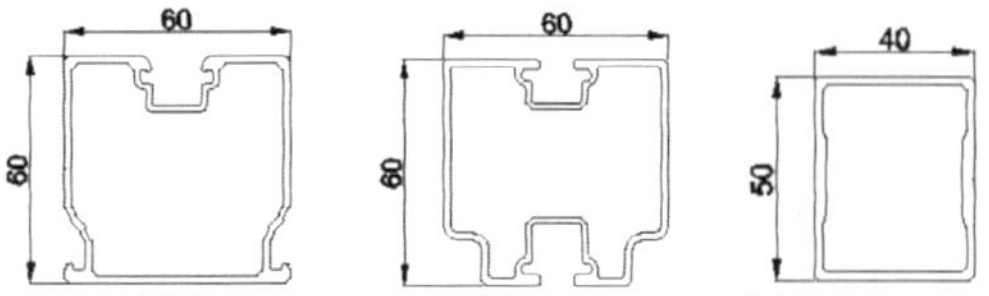

(a)Rail (b)Beam (b)Column

Figure 1. Section of photovoltaic support profiles

2.3 Method of simulation

(1) According to the design guide on structures for photovoltaic array (Japanese Industrial Standard, JIS C 8955-2011), combined with the basic parameters of the project, the typical permanent load, wind load, snow load and seismic load of the photovoltaic support were determined.

(2) According to the two ends of the beam theory, combined with the beam number of designs, the influences of different quantities and span of beam on rail performance were discussed, so as to make sure the number and span of beam.

(3) With the single factor experiment, using SAP2000 software discussed the different support layout effects on the beam performance, so as to make sure the best supporting point position of beam.

(4) When the overall structure scheme was determined, the SAP2000 software was used for further discussion of the effects of different section thickness on the performance of rail and beam. At the same time, the strength of column and bolt was checked, and it provided the basis for the lightweight photovoltaic support structure design.

3 Analysis and discussion of results

3.1 Analysis and determination of PV stent load

The load of photovoltaic stent consisted of two parts, the permanent load and live load respectively. The permanent load (G) included the weight of photovoltaic module (G_1), rail weight (G_2), beam weight (G_3). The live load included wind load (W), snow load (S), and seismic load (K), and the load combination classification was shown in table 2 [8-11].

Table 2. Classification of load combinations

Load condition		Distinguishing	
		General area	Snowy area
Long-term	The usual	G	$G+0.7S$
Short-term	Snow	$G+S$	$G+S$
	Storm	$G+W$	$G+W$
			$G+0.35S+W$
	Earthquake	$G+K$	$G+0.35S+K$

The permanent load was mainly composed of the weight of PV module, rail and beam and the thickness of each was 2 *mm*. The total load was set as follow.

$$G=G_1+G_2+G_3 \quad (3\text{-}1)$$

Wind load W was related to wind force coefficient C_w, design speed pressure q_p and the area of modules A_w. In actual situation, the PV modules were divided into two cases: downwind load (W_+) and upwind load (W_-). By calculation, two kinds of wind load value could be gotten.

$$W_+=C_{w+}\times q_p\times A_W \quad (3\text{-}2)$$

$$W_-=C_{w-}\times(-q_p)\times A_W \quad (3\text{-}3)$$

Snow load was determined by the average unit load of snow P, vertical snow cover Z_s, snow area A_s and slope coefficient C_s. The snow load value was as follow .

$$S=C_s\times P\times Z_s\times A_s \qquad (3\text{-}4)$$

This project belonged to the general area, so the calculated seismic load was as follow.

$$K=k_p\times G \qquad (3\text{-}5)$$

In which, k_p was the design of horizontal seismic degree.

After finding out the permanent load, wind load, snow load and seismic load, the maximum load combination could be obtained according to the combination of the general areas in Table 2.

$$P_{max}=G+W=G+|W_+|\times cos\alpha \qquad (3\text{-}6)$$

Therefore, the uniformly distributed load on the rail was as follow.

$$q=P_{max}/((n-1)\times L) \qquad (3\text{-}7)$$

Similarly, the maximum horizontal load combination was as follow.

$$P_x=|W_-|\times sin\alpha \qquad (3\text{-}8)$$

The maximum combination of load in vertical direction was as follow.

$$P_z=G+|W_+|\times cos\alpha \qquad (3\text{-}9)$$

Among them, N - rail number, L - rail length (mm).

In summary, according to the basic parameters provided by the project, combined with the above formula and the design guide on structures for photovoltaic array JIS C 8955-2011, the calculation results were shown in table 3.

Table 3. Key parameters of the photovoltaic stent load

Name	Code	Parameter (N)
Permanent load	G	4679.4
Downwind load	W_+	30925.5
Upwind load	W_-	-37426
Snow load	S	27933.6
Earthquake load	K	5877.51
Maximum load combination combination	P_{max}	36244.8
Maximum load in horizontal direction	P_x	9686.5
Maximum load in vertical direction	P_z	35749.3

3.2 Optimization of beam span

The known PV module specification was 1650mm× 991mm× 40mm, and the modules were arranged in 4 rows and 5 columns. Each module had a horizontal distance of 10 mm, and the rail end was 50 mm away from the edge of the module, so the rail length was 5×1650+50×2+10×4=8390 mm. According to the two ends extended beam model of material mechanics (Figure 2 (a))[12], the PV support force was simplified into uniform load. The photovoltaic stent could be simplified into a multi fulcrum supporting beam whose ends were extended (Figure 2 (b)). The rail was divided into 2 sections l_1 and N of l_2, and the model was symmetrical distribution.

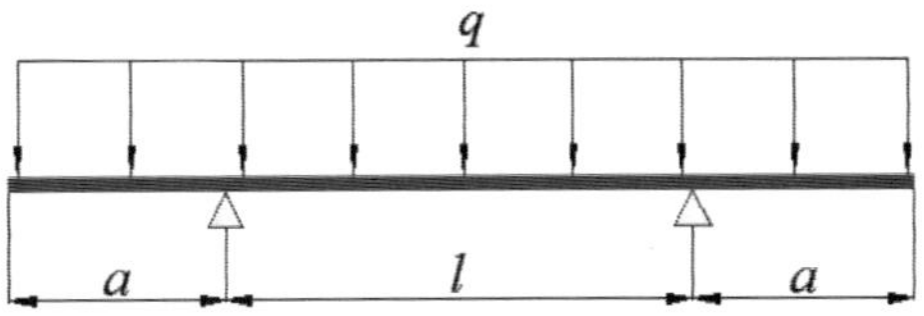

(a) The two ends extended beam model of material mechanics

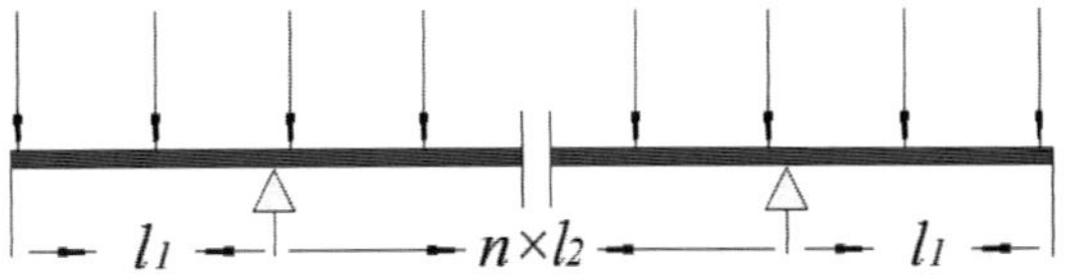

(b)Simplified model of rail with multi fulcrum supporting

Figure 2. Stress model

According to the maximum deflection formula of static perturbation based on the two ends extended beam model[13], if the deformation of the two ends extended beam model was minimum, it could be seen that the relationship between the parameters a and l could be obtained just only when the W_{max} was zero, here $l\approx 2.19a$.

$$W_{max}=\frac{ql^4}{384EI}\left(5-24\frac{a^2}{l^2}\right) \qquad (3\text{-}10)$$

Based on the above relationship, when deformation of l_1 was equal to l_2, the stent deformation tended to be coordinated, and it would not appear a part of the deformation which was too large to avoid increasing the section size or thickness, saving materials and reducing cost. According to the relationship between a and l when the deformation was minimum of the two ends of the beam model, the specific parameters of the number of different span beam had been discussed, as shown in table 4.

Table 4. The number of the corresponding parameters of different span beam

Number of oblique beam	l_1 *(mm)*	l_2 *(mm)*
2	2002	4385
3	1315	2880
4	979	2144
5	780	1708
6	648	1419

When the number of beam was greater than 4, the span of rail was too small and wasted materials; when the beam number was 3, the length of cantilever was too large, close to a single block's length; when the number of beam was 1, the span of rail and cantilever were too large. In summary, the number of beam was 4, and the rail span and cantilever were 2200 *mm*, 895 *mm*, as shown in figure 3.

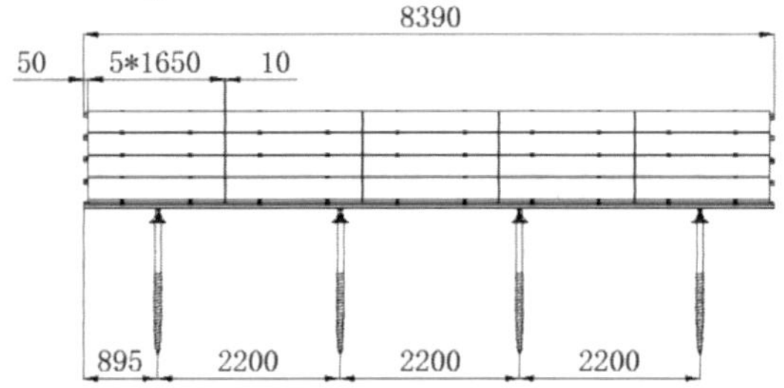

Figure 3. Arrangement of photovoltaic stents

3.3 Optimization of the position of connection point between beam and column

The known PV module specification was 1650*mm*× 991*mm*× 40*mm*, and the modules were arranged in 4 rows and 5 columns. Each module was longitudinally provided with a pressing block with a 23.5 *mm* distance and with a 50 *mm* from the end of the beam module edge, so the beam length was 4× 991 + 100× 2 + 23.5× 3 = 4234.5 *mm*. The length was 4300 mm after rounded. The support spacing between beam and pillar was determined by single factor experimental method. With six sets of data, the distance between the support point and the endpoint was 100 *mm*, 200 *mm*, 300 *mm*, 400 *mm*, 500 mm, 600 *mm* respectively, using SAP2000 software for simulation, as shown in Figure 4. The structure was symmetrical, so it just needed to monitor two points of contact deformation, and namely the monitoring points were P_1 and P_2 . And the two monitoring characteristic points were end point P_3 and span midpoint P_4 of beam respectively.

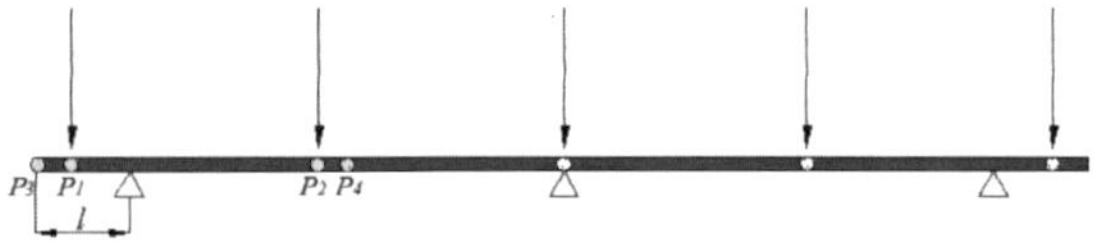

Figure 4. Analysis model of beam

In figure 4, the thickness of beam of the model was 2 *mm* and the concentrated load of each point was 500 *N*. According to the data obtained by SAP2000, statistical analysis was carried out. As shown in fig.5, the deformation tendency of the monitoring point was different when the supporting point spacing *l* was different. It was shown from the graph that the global average deformation was minimum at *l* = 400 *mm*, and the intersection point of deformation track of all monitoring points was between 400-450 *mm*, so the cantilever support position took 400 *mm*.

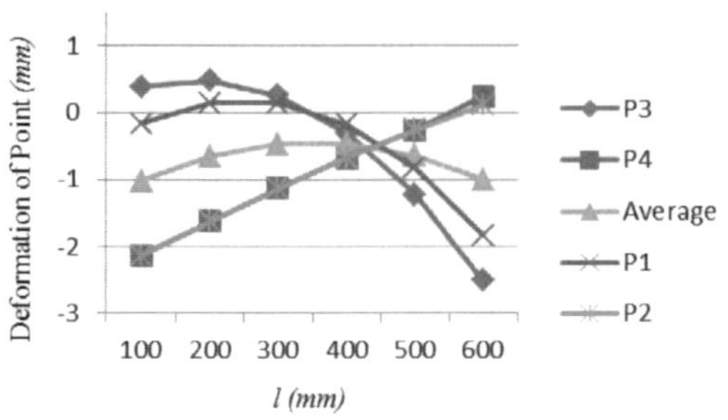

Figure 5. Deformation of monitoring points

3.4 Lightweight design of guide rail

On the one hand, the rail should meet the flexibility of the photovoltaic stent specifications (*l*/100, *l*-the length of rail's cantilever, 895 *mm*.), on the other hand, the strength of materials also should meet the requirements. And as known, the yield limit of aluminum profile σ_s was 240 *MPa* and the safety factor was 1.5, so the [σ] was as follow.

$$[\sigma]=\sigma_s/n=160\ MPa \qquad (3\text{-}11)$$

Among them, *N* - rail number, *L* - rail length (*mm*).

According to the engineering experience and the specification of the factory aluminum production, the maximum thickness of the aluminum sections was chosen to carry out the simulation. When the thickness of rail section was 2 *mm*, the deformation was 4.074 *mm* and the maximum stress was 44.697 *MPa* which met the requirements, and there was a large amount of surplus. Combined with the specific profiles of the company produced section, the calculation and analysis of deformation and the maximum stress value of different section thickness were carried out, then the thickness was optimized. The analysis results were shown in table 5.

Table 5. Statistics of deformation of rail with different thickness

Thickness	Deformation（*mm*）	Stress（*Mpa*）
1.0*mm*	5.801	62.595
1.2*mm*	5.317	57.63
1.4*mm*	4.93	53.655
1.6*mm*	4.602	50.244
1.8*mm*	4.320	47.285
2.0*mm*	4.074	44.697

From the change rule in table 5, the rail of different section thickness could meet the requirements of the company. From the view of lightweight and material saving, selecting the 1.0 *mm* thickness of rail was good. The maximum deformation was 5.801 *mm* < 895/100,

 https://doi.org/10.1051/matecconf/201816603002

and the stress value was 62.595 *MPa* < 160 *MPa*, which met the requirements.

3.5 The lightweight design of beam

Similarly, the beam also needed to meet the specification of the photovoltaic stents allowable deformation value ($l/200$, l = 1750) and material strength ([σ] =160 *MPa*) requirements like rail.

According to the results of section 3.1, and the number of the beam in section 3.2 was 4, with the module horizontally placed, the force on the supporting edge of the rail was half of the middle point. So, it would get the rail and beam contact stress formula (3-13) and (3-14).

$$F_X=P_X/((n-l)\times N)=605.4N \quad (3\text{-}13)$$

$$F_Z=P_Z/((n-l)\times N)=2234.3N \quad (3\text{-}14)$$

Among them, n——the number of rail, N——the number of beam.

Figure 6 was the force diagram of the beam structure of photovoltaic support.

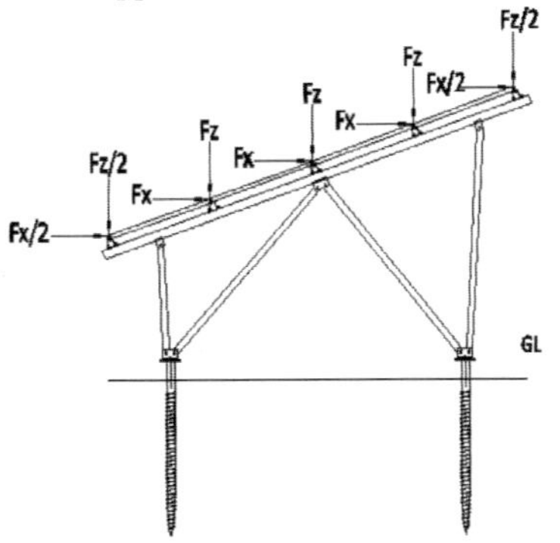

Figure 6. Force diagram of structure

Same as rail, when the thickness of rail section was 2 *mm*, the deformation was 4.688 *mm* and the maximum stress was 67.287 *MPa* which met the requirements, and there was a large amount of surplus. Combined with the production of company with specific section, the calculation and analysis of deformation and the maximum stress value of different section thickness were carried out, then the thickness was optimized. The analysis results were shown in table 6.

From the change rule in table 6, the rail of different section thickness could meet the requirement of the company. From the view of lightweight and material saving, selecting the 1.0 *mm* thickness of rail was good. The maximum deformation was 6.822 *mm* < 1750/200, and the stress value was 98.416 *MPa* < 160 *MPa*, which met the requirements.

3.6 Strength inspection of column

Rail and beam specifications had been determined, and the column specification was 40×50×2. In the structural model, each column was equivalent to two-force members, and only the axial force was required and checked. According to the force analysis results of the known model in SAP2000, the axial force of each column was shown in table 7.

Table 6. Statistics of beam deformation with different thickness

Thickness	Deformation（*mm*）	Stress（*Mpa*）
1.0*mm*	6.822	98.416
1.2*mm*	6.247	89.976
1.4*mm*	5.815	83.886
1.6*mm*	5.455	78.781
1.8*mm*	4.941	70.982
2.0*mm*	4.688	67.287

Table 7. Statistics of column axial force

Name	Axial force（*N*）
Front column 1	2449.63
Front column 2	1864.82
Rear column 1	4380.51
Rear column 2	2650.52

According to table 7, the maximum axial force of column was 4380.51 *N*, and the sectional area of column was A = 320.46 mm^2, so the maximum stress of the column was as follow [13,14].

$$\sigma=4380.51/A=4380.51/320.46Mpa=13.67Mpa<[\sigma] \quad (3\text{-}15)$$

Therefore, the strength requirements were met.

3.7 Bolt strength check

In the ground fixed support system, the connecting bolts bore the load in the form of shear between column and pile, pillar and beam, pile and beam. The maximum shear force was required to be checked, and it could be seen from table 6 that the value was 4380.51 *N*. The model adopted M10 bolt, whose material was Q235 ordinary carbon structural steel, and the yield limit was 215 *MPa*.

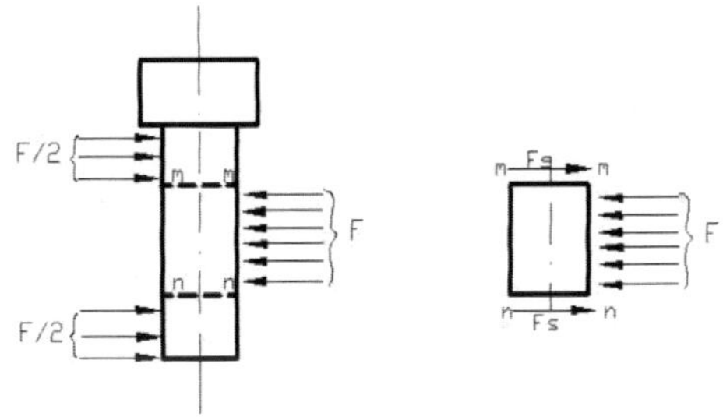

Figure 7. Shear model of bolt

The shear model of bolt was shown in figure 7, and F was the shear force of the bolt. The strength condition was as follow [15,16].

$$\frac{4F}{\pi d^2 m} \leq [\tau] \quad (3\text{-}16)$$

In which, d was diameter of shear surface of bolt; m was number of bolt shear planes; [τ] was allowable shear stress of bolt, and the safety factor was 2.5. The shear stress on a single section was as follow.

$$\frac{4F}{\pi d^2 m} = \frac{4\times4380.51}{\pi\times(10\times10^{-3})^2\times4} = 13.94Mpa<[\tau] \quad (3\text{-}17)$$

Therefore, the strength requirements were met.

4 Conclusion

In this paper, the load calculation, structural design and optimization, and the strength calculation of the connection parts of the fixed photovoltaic support system are carried out through the actual project case design, which are as follows:

(1)According to the general requirements of the 4 rows and 5 columns fixed photovoltaic stent, the length of the rail is 8390 *mm*. The initial support profiles thickness is 2 *mm*, and the typical permanent load of the photovoltaic support is 4679.4*N*, the wind load 1.05 kN/m^2, the snow load 0.89 kN/m^2 and the seismic load 5877.51*N*.

(2)By the two ends of the beam model of theoretical calculation, comparing the span of the rail and the length of the cantilever when the number of the beam is different, the beam number and rail span are ultimately determined 4 and 2200 *mm*.

(3) By using the single factor experiment method, through the calculation and analysis of SAP2000, the three best supporting positions of the W strut are determined.

(4) By the simulation, and the photovoltaic support design requirements, the optimal parameters are for the rail 60× 60× 1.0, beam 60× 60× 1.0, column 40× 50× 2, bolt M10.

Acknowledgements

This work was financially supported by A Priority Academic Program Development of Jiangsu Higher Education Institutions (PAPD), Jiangsu Policy Guidance Class Program (Industry- University- Research Collaboration) Project (BY2015047-02), Jiangsu Wall Material Innovation Research Project (201702), Frontier and Key Technical Innovation Program of Nantong (MS22015028), Open Fund Project of Jiangsu Key Laboratory of 3D Printing Equipment and Application Technology (Nantong Institute of Technology) (2017KFKT02 、 2017KFKT09 、 2017KFKT10), and Nantong Key Laboratory of 3D printing technology and Application (CP12016002).

References

1. A. Girard, E. J. Gago, J. Ordoñez, et al, *Renewable Energy*, **86**, 703 (2016).
2. E. Drury, A. Lopez, P. Denholm, et al, *Progress In Photovoltaics*,**22**(12), 1302 (2014).
3. H. Khonkar, A. Alyahya, M. Aljuwaied, et al, *Solar Energy*, **110**, 268 (2014).
4. R. R. Hernandez, M. K. Hoffacker, M. L. Murphy-Mariscal, et al, *Proceedings of the National Academy of Sciences*, **112** (44), 13579 (2015).
5. X. X. Wang, H. Y. Lu, H. J. Ni et al, *AER-Advance in Engineering Research*, **22**, 297 (2015).
6. H. J. NI, W. F. LU, X. X. Wang et al, *AER-Advance in Engineering Research*, **22**, 302 (2015).
7. C. K. Lin, C. Y. Dai, Wu J C, *ReneWable Energy*, **59**(6), 65 (2013).
8. Japanese Industrial Standard, *JIS C 8955-2011 Design guide on structures for photovoltaic array* (2011).
9. H. Huang, M. Zhang, Y. T. He. *Solar Energy, Chinese*, 36(6), 47(2015).
10. B. García-Domingo, M. Torres-Ramírez, J. D. L. Casa, et al, *Energy & Buildings*, **83**, 130 (2014).
11. Beijing Jintuwu Software Technology Co., Ltd. China Institute of Building Standarddesign & Research, *Guidelines of SAP2000 in Chinese* (Beijing: China Communications Press, 2006).
12. Z X. Guo, *Static Caculation Handbook for Practical Structure* (Beijing: China Machine Press, 2009).
13. H. W. Liu, Static Caculation Handbook for Practical Structure (*Beijing: Higher Education Press*, 2011).
14. D. X. Cheng, *Machine Design Hand Book* (Beijing: China Machine Press, 2007)
15. H. Q. Lu, *Proceedings of Chinese society for* electrical *engineering the fifth conference on civil engineering branch of structure* (Nanning, China, 2006).
16. Chinese national standards, *GB 50797-2012 Design Code for Photovoltaic Power Station* (2012).

A Comparative Study of Refrigerant Side Condensing Heat Transfer Coefficient for Horizontal Mini Channel

Ashok Parekh[1], Tejendra Patel[2], Parbhubhai Tailor[3]

[1]*Associate professor, MED, SVNIT Surat, India.*
[2]*Research Scholar, MED, SVNIT Surat, India.*
[3]*Professor, MED, SVNIT Surat, India.*

Abstract. Heat exchanger developments are driven by increase in energetic efficiency and emission reduction. To reach the required standards, new systems are required based on mini-channels along with macro channels also. Mini-channels can be described as tubes with one or more ports extruded in aluminium or any other material like copper, steel etc. with hydraulic diameter in the range of 0.2 to 3 mm. Its use in refrigeration systems became a reality; thanks to the human ability to make micro-scale systems. Some heat exchanger enterprises have some models developed specially for their use in automotive sector, cooling sector and industrial refrigeration without having knowledge of how these reduced geometries affect the most important parameters such as pressure drop and the heat transfer coefficient. To overcome this, an exhaustive literature review of the last two decades has been performed to determinate the state of the research. Among all the publications, several models have been reviewed to check the predicting capacities of them as most of them were developed for macro channels along with single port mini-channel tubes. This work presents a comparative study of existing heat transfer correlations for horizontal flow through mini-channels. Present work total 317 experimental results of condensation heat transfer coefficient obtained from several published literatures and also various condensation heat transfer correlations are reviewed for horizontal flow through mini-channels. Results of condensation heat transfer coefficient obtained for various correlations are compared with the experimental results. Among all correlations used the Bohdalet's correlation gives value of condensation heat transfer coefficient closer to experimental results and has a mean relative deviation (MRD) of 5.4 % whereas mean absolute relative deviation (MARD) of 20.72%.Hence in present work the effort has been made to find better model to predict the condensation heat transfer coefficient which can fit the experimental data with least MRD and MARD.

1 Introduction

Nowadays, micro and mini-channels are present in many applications ranging from different heat exchangers in process industry to automotive, electronics and domestic applications. Two-phase flow has been applied to a growing number of fields in recent years because of its higher energy efficiency in comparison with single-phase flow. Compactness is a synonym of charge reduction and this is very important in present day refrigeration systems and heat pumps because of the great contribution of HCFC and HFC refrigerants to the direct greenhouse effect. This reduction is also important for natural refrigerants such as hydrocarbons and ammonia for safety reasons [1]. One of the pioneers authors who studied the influence of reducing diameter in heat transfer coefficient were Kays and London [2]. From then on, the investigation on heat transfer in mini-channels has been one of the most researched topics in this field. Even though many researchers suggested various correlations, they all show a considerable deviation for many new refrigerants. It is partly due to the fact that most of the correlations were based upon a set of data for a few specific fluids and hence certain governing factors valid only for those fluids were mainly emphasized in those correlations. Therefore, it is necessary to develop a correlation based upon a variety of data of many fluids including new alternative refrigerants measured from the same apparatus by a consistent method. Thome et al. [3] correlation presented the lower deviation, less than 30%.Al-Hajri et al. [4] experimentally studied two-phase condensing flows of R134a and R245fa in a single mini-channel of 0.7 mm diameter with high aspect ratio. Pressure drop is accurately predicted by Lockhart-Martinelli with deviation lower than 20 %. Heat transfer coefficient is predicted with errors lower to 20 % by Dobson and Chato correlation with a modified power of the Martinelli parameter. Heo et al. [5] reported a study about in-tube condensation heat transfer characteristics of CO_2 in different mini-channels. Multi-port mini channels had hydraulic diameters of 1.5, 0.78 and 0.68 mm and they were tested from -5 to 5 °C of saturation temperature.

Wang et al. [6] presented a short overview of the heat transfer performance of R1234yf versus R134a with previous published data. For in-tube condensation, it was found that the condensation heat transfer coefficients for R1234yf are inferior to those of R134a. Flow pattern influence is very strong in heat transfer processes and momentum. A good revision of recent condensation heat transfer coefficient and pressure drop in tubes of small and large diameter can be found in English and Kandlikar [7].Yang and Webb [8] studied R12 condensation in plain micro-finned multi-port extruded tubes with hydraulic diameters of 2.637 and 1.564 mm. They compared against the correlations developed by Akers et al. [9] and Shah and London [10]. They found that Akers et al. [9] correlation predicted well their experimental data for plain tubes. The authors also showed that high velocity data are over estimated. Yang and Webb et. al. proposed a correlation for heat transfer coefficient prediction in finned mini-channels. This correlation takes into account vapour shear tension and surface tension effects. They proved that at high mass velocities, heat transfer coefficient is controlled by vapour shear tension and the contribution of surface tension is quite small (it is much affected by fin geometry). These authors also proposed a separated asymptotic model for these two effects. They validated the previous model analyzing five different geometries (a single port tube and several multi-port tubes with and without fins) and two refrigerants, R12 and R134a.Webb et al. [11] studied a circular copper tube with a hydraulic diameter of 3.55 mm and an aluminium multi-port tube of 2.13 mm. They checked their experimental results for R134a with those provided by Shah M.M. et al. [12] and Moser et al. [13] correlations under similar conditions. They found good concordance between their data and both correlation predictions, at 65 °C saturation temperature; they showed that Shah M.M. et al. correlation overestimates their data. The authors attributed this difference to the reduced pressures. In their data, it was $p_{red} = 0.47$ while Shah M.M. et al. correlation was developed for $0.011 < p_{red} < 0.44$. Accordingly, round single port tubes and multi-port tubes are going to reach the same condensation performance.

2 Comparative Study of CHTC for Condensation in Annular Flow Regime

Here, nearly about 209 of experimental data have been taken from literature (Experimental measurements of Aker's et al. , A. Cavallinia at al [14], Dongsoo Jung at al [15] obtained for a 0.952 mm and 1.304 mm square channel whereas 1.152 mm circular minichannel at different saturation temperature) for different refrigerant as working fluid. Comparative assessment of six available different correlations have been done and finding MRD (Mean Relative Deviation) of each correlation. Generally, annular flow correlations are classified into three categories, viz. two-phase multiplier based, shear-based and boundary layer based as given by Dobson et al. [16].The characteristics of the heat transfer mechanism for forced convection condensation can be explained by the Akers etal. [17] correlation, presented in Eq. (1), which was developed for annular flow regime.

$$Nu = 0.026Re_L^{0.8}Pr_L^{0.333}((\frac{x}{1-x})(\frac{\rho_L}{\rho_V})^{0.5} + 1)^{0.8} \quad (1)$$

Cavallini et al. [18] proposed a HTC for film condensation inside horizontal tubes.

$$Nu = 0.05Re_L^{0.8}Pr_L^{0.333}((\frac{x}{1-x})(\frac{\rho_L}{\rho_V})^{0.5} + 1)^{0.8} \quad (2)$$

The Shah et. al. correlation has been compared by researchers commonly for turbulent condensation conditions valid for $Re_l \geq 350$.

$$Nu = 0.023Re_{Lo}^{0.8}Pr_L^{0.4}\{1 + 3.8[\frac{x/(1-x)]^{0.76}}{p_{rd}^{0.38}}\}(1-x)^{0.8} \quad (3)$$

Chen et al. [18] developed a generalized correlation for vertical flow using the pressure drop model of Dukler in 1960 and neglecting acceleration head.

$$Nu = 0.018\left(\frac{\rho_L}{\rho_V}\right)^{0.39}\left(\frac{\mu_v}{\mu_L}\right)^{0.078}Re_L^{0.2}(Re_{Lo} - Re_L)^{0.7}Pr_L^{0.65} \quad (4)$$

Dobson and Chato et. al. developed a correlation using a two-phase multiplier for an annular flow regime. It is recommended for G>500 kg/m^{2}s for all qualities in horizontal tubes.

$$Nu = 0.023Re_L^{0.8}Pr_L^{0.3}(\frac{2.61}{X_u^{0.805}}) \quad (5)$$

$$Nu = 0.0274Pr_L Re_L^{0.6792}x^{0.2268}\emptyset_v/x_n \quad (6)$$

$$\emptyset_v^2 = 1.376 + 8X_u^{1.665} \quad (7)$$

$$\mathrm{Nu} = \frac{\mathrm{hd_h}}{\mathrm{k_L}} \quad (8)$$

$$\mathrm{Re_{Lo}} = \frac{\mathrm{Gd_h}}{\mu_L} \quad (9)$$

$$\mathrm{X_u} = \left(\frac{\mu_L}{\mu_V}\right)^{0.1}\left[\frac{1-x}{x}\right]^{0.9}\left(\frac{\rho_L}{\rho_V}\right)^{0.5} \quad (10)$$

$$\mathrm{p_{rd}} = \frac{\mathrm{p_{saturted}}}{\mathrm{p_{critical}}} \quad (11)$$

Bohdal et. Al [23] correlation

$$Nu = 25.08Re_L^{0.258}Pr_L^{-0.495}P_{rd}^{-0.288}[\frac{x}{1-x}]^{0.266} \quad (12)$$

3 Comparative Study of CHTC for Film Condensation

Here, near about 108 of experimental data have been taken from literature (Experimental measurements of Matkovic et al. [19], A. Cavallinia at al [20], Dongsoo Jung at al [21] obtained in a 1 mm internal hydraulic diameter circular mini channel at 40 ^{0}C saturation temperature) for R134a refrigerant as working fluid. Comparative assessment of four available different correlations have been done and finding MRD (Mean Relative Deviation) of each correlation. After comparison it has been found that the Bohdal et. al. [22] correlation and haraguchi et al. [23] correlation having minimum MARD (Mean absoluteRelative Deviation) 20.72 % and 20.75 % respectively. It means predicted heat transfer coefficient by using haraguchi's correlation fits the experimental data of Matkovic et al., A.Cavallinia et al and Dongsoo Jung et al. The correlations being compared includes the Shah correlation, the Cavallini correlation, the Aker's correlation, the huang's correlation, the Koyama's correlation and the Haraguchi correlation, as summarized in Table 2 shows mean deviation and accuracy of the nine correlations for the condensation

Table 1. Experimental data sources for condensation heat transfer

Data Sources	Refreigerant	Parameter Range: Tsat(^{0}C)/ G(kg/m^{2}s)/x	Geometry Range: D$_h$(mm)/L(mm) /Orientation	Number of Data points
Matkovic et al. [21] Cavallini a, et al.[18]	R134a	40 ^{0}C/65-1000/0.098-0..99	1/350/circulalar mini-channel Horizontal flow	108
Cavalliniaat al [18] Dongsoo Jung at al [20]	R22/R32/R152A	40^{0}C/150- 280/0.097-0.70	0.952-1.304/380/square & circular mini-channel Horizontal flow	209

Table 2. Comparison between experimental data and correlation prediction

Data Sources	Errors %	Corrélations								
		Akers	Cavallini	Shah	Chen	wang	Harahuchi	Dobson &Chato	Koyama	Bohdal
Cavallinia at al. [18]	MRD	-36.49	21.73	-54.23	43.22	-26.69	---	----	---	-12.262
	MARD	36.49	41.035	54.23	59.052	30.98	---	----	---	17.875
Dongsoo Jung at al. [20]	MRD	-25.47	42.93	-35.78	58.99	-21.38	---	----	---	23.15
	MARD	25.47	46.92	35.78	61.38	27.70	---	----	---	23.57
Matkovic et al. [21]	MRD	18.65	-17.56	20.034	--	--	-10.47	-24.31	-19.89	--
Cavallinia et al.[18]	MARD	28.65	20.56	24.34	---	--	20.75	25.14	29.85	---
Average	**MRD**	**-14.43**	**15.7**	**-23.32**	**51.10**	**-24.03**	**-10.47**	**-24.31**	**-19.89**	**5.4**
	MARD	**30.20**	**36.18**	**38.11**	**60**	**29.34**	**20.75**	**25.14**	**29.85**	**20.72**

heat transfer coefficient. Researchers have proposed number of their developed correlation to predict the heat transfer of pure substances and mixture boiling and condensing in or on horizontal and vertical tubes, or on vertical plates.

Total 317 experimental data points have been taken from the sources as presented in Table I are used for the comparative study of the six heat transfer correlations as described in section 2. The results obtained are presented in Table 2, where the MARD is the mean absolute relative deviation and MRD is the mean relative deviation.

Harahauchi's correlation [22]

$$h_{pridicted} = 0.0152 \times \frac{k_l}{d_h} \times (1 + 0.6Pr_l^{0.8}) \times \left(\frac{\phi_g}{\chi_{tt}}\right) \times Re_l^{0.77} \quad (13)$$

Where

$$\phi_g = 1 + 0.5 \times \left(\frac{G}{\sqrt{g d_h \rho_g (\rho_l - \rho_g)}}\right)^{0.75} \times \chi_{tt}^{0.35} \quad (14)$$

Koyama's correlation

$$h_{pridicted} = 0.0152 \times \frac{k_l}{d_h} \times (1 + 0.6Pr_l^{0.8}) \times (\frac{\phi_g}{\chi_{tt}}) \times Re_l^{0.77} \quad (15)$$

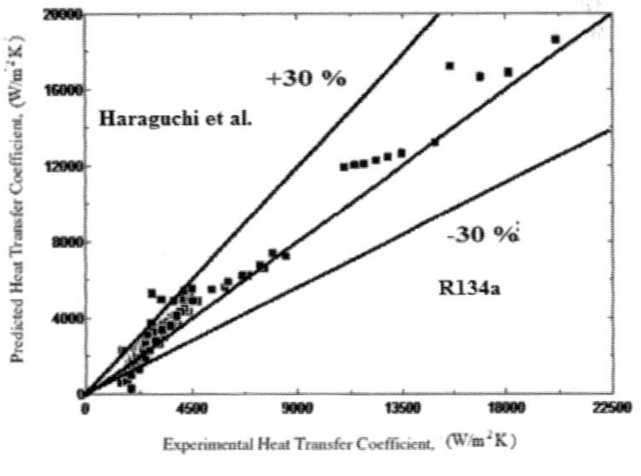

Figure 1. Comparison between experimental and predicted heat transfer coefficient using Haraguchi correlation.

Where

$$\phi_g = 1 + 21 \times (1 - e^{-0.319\, d_h}) \times \chi_{tt} + \chi_{tt}^2 \quad (16)$$

$$MRD = \frac{1}{N}\sum_{i=1}^{N} \frac{h(i)_{cal} - h(i)_{exp}}{h(i)_{exp}} \quad (17)$$

$$MARD = \frac{1}{N}\sum_{i=1}^{N} \left|\frac{h(i)_{cal} - h(i)_{exp}}{h(i)_{exp}}\right| \quad (18)$$

where, h_{cal} is the calculated value, h_{exp} is the experimental value, and N is the number of data points.

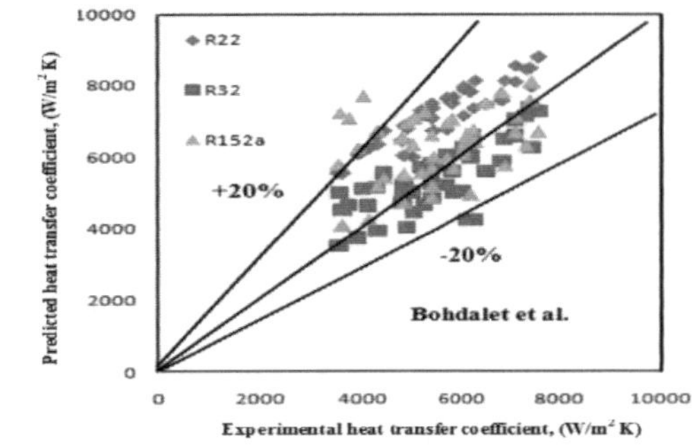

Figure 2. Comparison between experimental and predicted heat transfer coefficient using Bohdal et al. correlation.

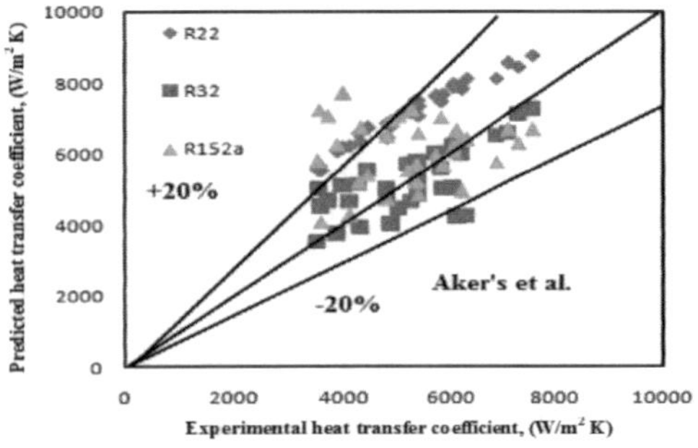

Figure 3. Comparison between experimental and predicted heat transfer coefficient using Akers et al. correlation.

The heat transfer coefficients are compared with the correlation prediction in Fig.1, Fig.2 and Fig. 3. For the annular flow regime based experimental HTC data of R32, R22 & R152a in square & circular mini-channel, the Bohdal et al. correlation predicts the data with a MARD 20.72 %. The MARD for the Akers et al. correlation is 30.20 %. For the film condensation experimental HTC data of R134a in circular mini-channel, the Haraguchis et al. correlation predicts the data with a MARD 20.75 %. As per past study of Travis et al. [24] the turbulent liquid film theory and Nusselt`s theory which somewhat dictates flow physics behind good agreement; they proposed an empirical asymptotic equation with the power of 2 for predicting the local Nusselt number. These models i.e. Haraguchi, Bohdalet and the Aker's correlation also takes into account free and forced convection. In addition to that both are annular flow regime based model or correlation and as per the literature past flow regime map study of Coleman & Gerimella [25] the flow pattern is expected to be annular flow or annular film and mist flow for most of the channel length and vapour quality range of $x > 0.7$. In the present study also there is all data for annular flow regime and film condensation within this range. Also the thermo physical properties of R134a, R32, and R22 & R152a do not show the adverse effect on these correlations.

4 Conclusions

Total 108 data points of film condensation heat transfer other than annular flow and 209 data point of condensation in annular flow regime are obtained from seven published papers and nine condensation heat transfer correlation are reviewed for flow inside smooth horizontal tubes. After analysis The comparisons show that the Bohdal et al. correlation and the Aker et. al. correlation have the best predictability of experimental data for horizontal flow inside smooth tubes with MRD 5.4 % and MRD -14.43 % respectively. Also it is seen that Haraguchi et al. correlation showing remarkable agreement with experimental data with MRD -10.47%.

Nomenclature

A	area, m^2
C_P	Specific heat at constant pressure, kJ/kgK
C_V	Specific heat at constant volume, kJ/Kg K
D	Smallest hydraulic channel diameter, mm
Fr	Froude number
g	Gravitational Acceleration, m/s^2
G	Mass Velocity, kg/m^2s
h	Heat transfer coefficient, W/m^2K
h_I	Heat transfer coefficient for turbulence region, W/m^2K
h_{lv}	Heat of vaporization, kJ/kg
I	Current, amp
k	Thermal conductivity, W/m K
L	Mini channel Length, m
Nu	Nusselt Number
P	Pressure, bar
Pr	Prandtl Number
Pr_w	Prandtl Number at wall
P_{red}	Reduced pressure
Q_{in}	Heat input, Watt
Q	Heat flux, kW/m^2
Re	Reynolds number
T	Temperature, °C
T_{sat}	Saturated temperature, °C
T_{wall}	Wall temperature, °C
t	Thickness, mm
Xtt	Martinelli parameter
x	vapour quality

Abbreviations

CHTC	Condensation heat transfer coefficient
CFCs	Chlorofluorocarbons
CHE	Compact Heat Exchanger
CB	coalescing bubble
GWP	Global warming potential
HTC	Heat Transfer Coefficient
HT	Heat transfer
HCFC	hydrochlorofluorocarbans
HFC	hydrofluorocarbans
MRD	Mean relative Deviation
MARD	Mean Absolute Relative Deviation
ODP	Ozone depletion potential
RMSE	Root mean square error
VCR	Vapour compression refrigeration

Greek Symbols

f	friction factor
σ	surface tension, N/m
τ	shear stress, N/m^2
μ	viscosity of condensate film, N s/m^2
ρ	density of condensate film, kg/m^3
χtt	Lockhart-Martinelli parameter
φ	vapor two phase pressure drop multiplier
α	void fraction or Volume fraction

References

1. F. Poggi, H. Macchi-Tejeda, D. Leducq, and A. Bontemps, Refrigerant charge in refrigerating systems and strategies of charge reduction, International Journal of Refrigeration, 31353-370, (2008).
2. W. M. Kays and A. L. London. Compact Heat Exchangers. McGraw-Hill. New York. Second Edition. (1964).
3. J.R. Thome, S. Nebuloni, "Numerical modeling of laminar annular film condensation for different channel shapes", International Journal of Heat and Mass Transfer 53 (2010) 2615–2627.
4. E. Al-Hajri, A.H. Shooshtari, S. Dessiatoun, and M.M. Ohadi, Performance characterization of R134a and R245fa in a high aspect ratio micro channel condenser, International Journal of Refrigeration, 36 588-600 (2013).
5. J. Heo, H. Park, and R. Yun, Condensation heat transfer and pressure drop characteristics of CO2 in a micro channel, International Journal of Refrigeration, 36 1657-1668 (2013).
6. C.C. Wang, An overview for the heat transfer performance of HFO-1234yf, Renewable and Sustainable Energy Reviews, 19 444-453 (2013).
7. S.G. Kandlikar, N.J. English , An Experimental Investigation into the Effect of Surfactants on Air-Water Two-Phase Flow in Mini channels, Heat Transfer Engineering, 27 99-109 (2006).
8. C.Y. Yang, R.L. Webb, Condensation of R-12 in small hydraulic diameter extruded aluminium tubes with and without micro-fins, International Journal of Heat and Mass Transfer, 39 791-800 (1996).
9. W.W. Akers, H.A. Deans, and O.K. Crosser, Condensing heat transfer within horizontal tubes, Chem. Eng. Progr., 54 (1958).
10. R.K. Shah, Laminar Flow Forced Convection in Ducts, Advances in Heat Transfer, (1978).
11. R.L. Webb, M. Zhang, and R. Narayanamurthy, Condensation heat transfer in small diameter tubes, Heat Transfer, 14 403-408 (1998).
12. Shah, M. M. "A General Correlation for Heat Transfer During film Condensation inside Pipes". Int. J. Heat and Mass Transfer, 22(4), 547–556 (1979).
13. K.W. Moser, B. Na, and R.L. Webb, A New Equivalent Reynolds Number Model for Condensation in Smooth Tubes, Journal of Heat Transfer, 120 410-417 (1998).
14. A. Cavallini a,*, G. Censi a, D. Del Col a, L. Dorettia,G.A. Longo b, L. ossetto , "Experimental investigation on condensation heat transfer and pressure drop of new HFC refrigerants (R134a, R125, R32, R410A , R236ea) in a horizontal smooth tube", International Journal of Refrigeration 24 73-87 (2001).
15. Dongsoo Jung*, Kil-hong Song, Youngmok Cho, Sin-jong Kim, "Flow condensation heat transfer coefficients of pure refrigerants", International Journal of Refrigeration 26, 4–11 (2003).
16. Dobson, M. K., & Chato, J. C. "Condensation in Smooth Horizontal Tubes". J. Heat Transfer, 120 (1), 193–213 (1998).
17. Akers, W. W., Deans, H. A., & Crosser, O. K., "Condensing Heat Transfer within Horizontal Tubes", Chemical Engineering Progress Symposium Series, 55,171–176 (1959).
18. Chen, S. L., Gerner, F. M., & Tien, C.L., "General film Condensation Correlation" Experimental Heat Transfer, 193–107 (1987).
19. Matkovic, A. Cavallini, D. Del Col, L. Rossetto, Experimental study on condensation heat transfer inside a single circular mini channel, Int. J. Heat Mass Transfer 52 (2009) 2311–2323.
20. Cavallini a,, G. Censi a, D. Del Col a, L. Doretti a,G.A. Longo b, L. ossetto , "Experimental investigation on condensation heat transfer and pressure drop of new HFC refrigerants (R134a, R125, R32, R410A, R236ea) in a horizontal smooth tube", International Journal of Refrigeration 24 73-87 (2001).
21. Dongsoo Jung*, Kil-hong Song, Youngmok Cho, Sin-jong Kim, "Flow condensation heat transfer coefficients of pure refrigerants", International Journal of Refrigeration 26 4–11(2003).
22. Bohdal T., Charun H., Sikora M.: Comparative investigations of the condensation of R134a and R404A refrigerants in pipe minichannels. Int. J. Heat Mass Transfer 54(2011), 1963–1974.
23. Haraguchi, H., Koyama, S., Fujii, T., "Condensation of refrigerants HCFC 22, HFC 134a and HCFC 123 in a horizontal smooth tube", (2nd report, proposals of empirical expressions for local heat transfer coefficient). Trans. JSME 60 245-252 (1994).
24. D.P. Traviss, W.M. Rohsenow, and A.B. Baron, Forced Convection Condensation in tubes: A heat Transfer Correlation for condenser design, ASHRAE Transactions, 79 157-165(1973).Coleman J.
25. W., Garimella S., "Two phase flow regime transition in microchannel tube: the effect of hydraulic diameter, orlando, ASME, Heat transfer devision, 71-83 (2000).

The Strength Analysis of CFM56 Engine Blade

Zhenzhen Liu[1], Zhixiong Chen[2], Jin Chen[1]

[1] *Nanchang Hangkong University School of Aircraft Engineering,student,2017, China*
[2] *Shanghai University of Engineering Science School of Automotive Engineering, associate professor,2017, China*

Abstract. Aero engine is a kind of thermodynamic machinery,which require have strict aerodynamic load, mechanical load and strong durability, its longevity depends largely on the life of its main components. In this paper, a series of studies are carried out on the strength of fan blades of CFM56 engine, which provide a reference value for improving the reliability of the engine. The finite element model of the engine fan is established by using CATIA's finite element software.The centrifugal stress distribution of the fan at different speeds and the influence of torque on fans under different speeds are calculated respectively, and the static strength of the fan is checked.

1 Introduction

The CFM56 is a high ratio bypass turbofan engine, which is based on F101 core technology, to meet the needs of the international military and civilian aircraft market in late 1980s, developed by CFM international (CFMI)company,which is composed by American GE Corporation and the French SNECMA company. Since its first model CFM56-2 acquired the airworthiness certificate in November 1979, by 2005, it has developed to 6 Series: CFM56-3, CFM56-5A, CFM56-5B, CFM56-5c, CFM56-7 and so on, a total of 28 models, whose thrust has covered 71~151 kN, and has become the driving force of 22 types of aircraft.By February 2005, the use of each type of CFM56 engine reached 14553, and the cumulative working time of the engine was 274511467 flight hours and 161778396 cycles. At present, there are 5796 aircraft powered by CFM56, with an average of one aircraft taking off every 4 seconds. At the same time, about 800 CFM56 engines are used in our country.

Aeroengine is the power device of an aircraft, and it is known as the "heart" of the aircraft, so the fan is one of the important components of turbofan engine, and its performance directly affects the performance of the whole engine. The air behind the fan is divided into two roads. One road is the outer culvert, and the other is connotation.

2 The stress calculation of fan blade

Because the structure, loading and restraint conditions of fan blades are all spatial properties, so it is necessary to do the finite element analysis of fan blades, before calculating the stress of fan blades under centrifugal load. The precondition of the elastic stress analysis of the blade line are as follows[1][2]:

(1)The deformation of the blade is small when working. In general, the size of the deformation is not predetermined in advance and can only be estimated by experience. From the existing engine, the fan blade has smaller deformation, which can be considered to meet the requirements of small deformation.

(2)The material of the blade has not yielded at work, that is, within the elastic range. At present, most of the engine blades are designed with elastic, which can meet this requirement.

(3)There are no contact between the blades or between the blades and other parts. So, the blades are discussed in this paper does not include the tenon part, and do not consider the adjacent crown and shoulder the contact vibration.

When the blade is discretized, the stress changes in a relatively large area, such as near the hub, more dense grid should be, and in the stress change of relatively small area, such as the near tip, can share some sparse grid, it should be noted that, when the unit are divided, the size of the unit's direction should be similar, and try to make the angle of each unit is small, so as to ensure that each unit has a good character.

2.1 Three dimensional fan model

According to the CFM56 engine, the fan model is set up in the 3D design software CATIA, as shown in figure 1.

The material used in the calculation is steel, and the material parameters are shown in table 1.

2.2 Finite element mesh of the fan

All the units are hexahedral units, and the size of the fan is 50mm. Because of the strength calculation of the fan blades, so the size of the blades of the mesh fan blades is 20 mm. The mesh is divided into 57403 units and 18459 nodes by using the mesh division tool in CATIA software,as shown in figure 2.

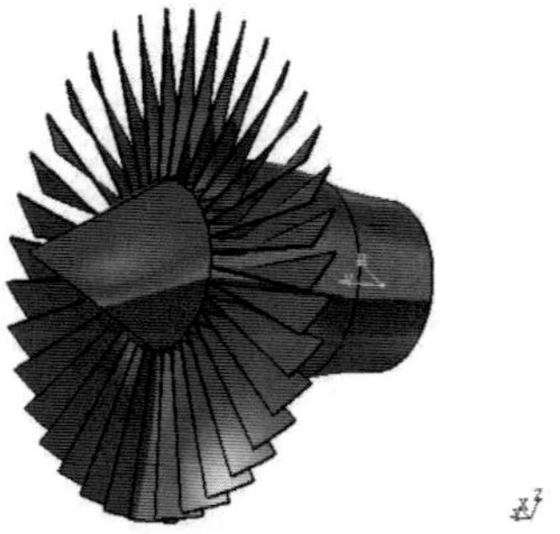

Figure 1. Fan model of CFM56 engine

Table 1. Physical parameters of steel

Material	Steel
Young's modulus	2E+011N·m^2
Poisson's ratio	0.266
Density	7860kg·m^3
Coefficient of thermal expansion	1.17E-005kdeg
Yield strength	2.5E+008N·m^2

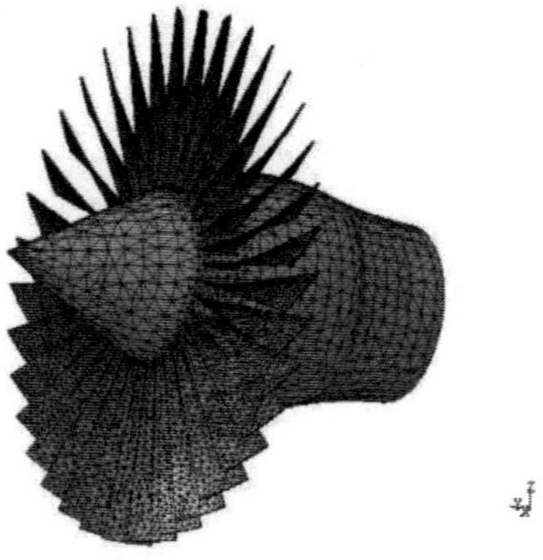

Figure 2. Finite element mesh of fan

2.3 Calculation of load and boundary conditions

2.3.1 Load characteristic of blade

The fan blades rotate around the engine axis at working, and the loads are included in the centrifugal force, aerodynamic force and thermal load.

The aerodynamic force is a kind of surface distribution pressure, it exerts effect on each surface of the blade, but it is not evenly distributed, its distribution along the leaf height direction and leaf width direction are not uniform,the aerodynamic field can be calculated accurately by the cascade passage and 3D flow field, but it is a larger workload. In general, aerodynamic forces should be taken into account when stress analysis is performed on the blade, but sometimes it can not be considered due to the lack of relevant data. The centrifugal stress on the fan blade is much larger than the aerodynamic stress,so ignoring the aerodynamic force has little effect on the analysis results.

Centrifugal force is a volumetric force, which is directly proportional to the mass, radius and the square of speed. It mainly causes the blade to generate radial tensile stress, which at the same time produces twisting stress for the twisted blade. The thermal load is generated when the blades are heated and the deformation of each part is restricted. Thermal stress generated by thermal load is not only related to the temperature gradient of the blade but also related to the geometric constraints on the blade. The larger the temperature gradient of the blade, the tighter the geometric constraint, the greater the thermal stress. Therefore, the thermal load is usually described by the temperature field and is closely related to the geometric boundary conditions. Strictly speaking, all the blades have the effect of thermal load, but for the fan blade, due to low temperature and uniform temperature distribution, and the thermal stress is not large, so the stress analysis is often neglected.

2.3.2 Analysis and calculation of centrifugal force

It is known by 2.3.1 that the stress analysis of the blade only needs to consider the influence of the centrifugal force. So this paper only calculates and analyzes the strength of the fan blade under the centrifugal load. The deformation and stress of the fan blade are calculated by the software CATIA.The rated speed of the fan is 14460 r/min, In order to analyze the stress distribution of fans at different speeds, three speeds of 8000 r/min, 14460 r/min and 20000 r/min are analyzed.The finite element software CATIA is used to analyze the stress at different speeds, as shown in Figure 3 to 4.

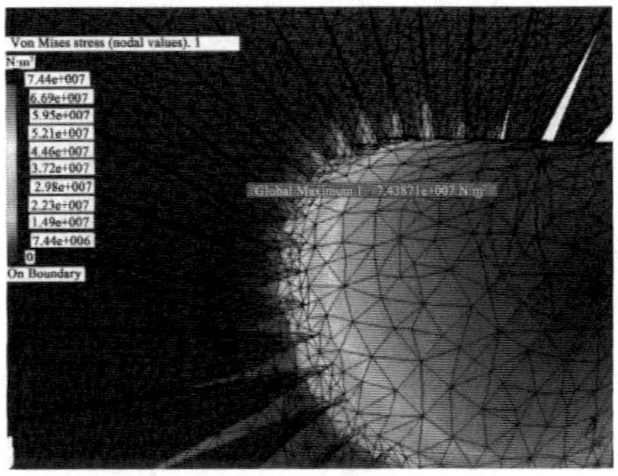

Figure 3. The stress distribution of fan at 8000r/min

From figure 3-4, we can see that when the speed is 8000 r/min, the stress of the fan is concentrated on the root of the blade, the maximum stress of the blade is 74.4MPa, the deformation displacement appears at the tip of the leaf, and the maximum radial displacement of the leaf tip is about 0.208mm.Likewise,when the speed is 14460 r/min, the stress of the fan is concentrated on the root of the blade, the maximum stress of the blade is 398.2MPa, the deformation displacement appears at the tip of the leaf, and the maximum radial displacement of the leaf tip is about 0.713mm. Similarly,when the speed is 20000 r/min, the stress of the fan is concentrated on the root of the blade, the maximum stress of the blade is 614.1MPa, the deformation displacement appears at the tip of the leaf, and the maximum radial displacement of the leaf tip is about 1.0948mm.

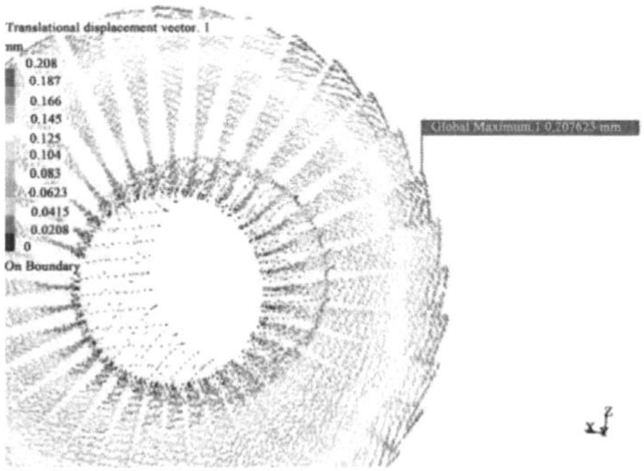

Figure 4. The stress deformation of fan at 8000r/min

From the above we can see that under different speeds, the fan's force is basically the same,the stress of the fan is concentrated on the root of the blade,the deformation displacement appears at the tip of the leaf. The greater the speed is, the greater the force of the blade is, the greater the radial displacement of the blade is.

2.3.3 The effect of the torque

Under the action of the centrifugal force field, the fan blade is not only stretched, but also bears the torsional torque to restore the torque. The larger the blade twist is, the greater the recovery torque will be. The torsional deformation of the blade will cause torsional shear stress on the cross section of the blade and redistribute the tensile stress on the cross section of the blade. In 2.3.2, the influence of centrifugal force is only considered.The influence of torque is not considered. The influence of torque on fan blades is analyzed separately here, and a torque is added to the fan blades.The specific method of calculating torque according to the rated speed of the fan is as follows:

$T=9550*P/n$

In the formula, T (N; m) is torque; P (kW) is power; the power in this article is 1800kW; n (r/min) is the speed, here the rated speed is 14460 r/min. The torque of the fan at the rated speed is 1189N • m. Three different torques are applied to the fan for analysis,and the stress diagram is shown in figure 5-6.

It can be seen from Figure 5-6 that when the torque is 800 N • m, the stress of the fan is concentrated at the root of the blade. The maximum stress of the blade is 4.32 MPa. The deformation displacement appears at the tip of the fan blade. The maximum radial displacement of the blade is about 0.127 mm.

Likewise, when the torque is 1189 N • m, the stress of the fan is concentrated at the root of the blade. The maximum stress of the blade is 6.4 MPa. The deformation displacement appears at the tip of the fan blade. The maximum radial displacement of the blade is about 0.327 mm.Similarly,when the torque is 1500 N • m, the stress of the fan is concentrated at the root of the blade. The maximum stress of the blade is 8.11 MPa. The deformation displacement appears at the 'tip of the fan blade. The maximum radial displacement of the blade is about 0.237 mm.

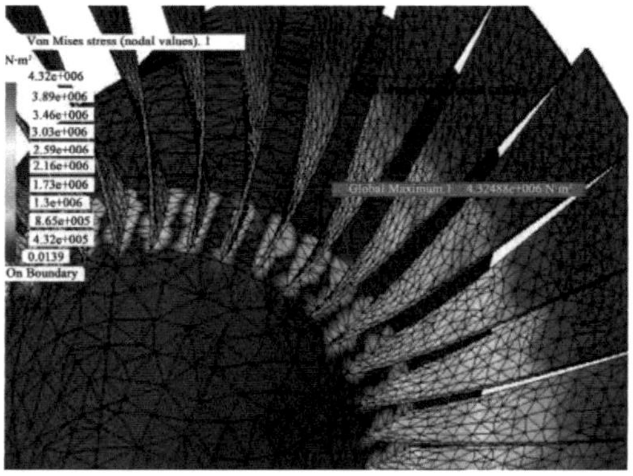

Figure 5. The stress distribution of fan at 800 N • m

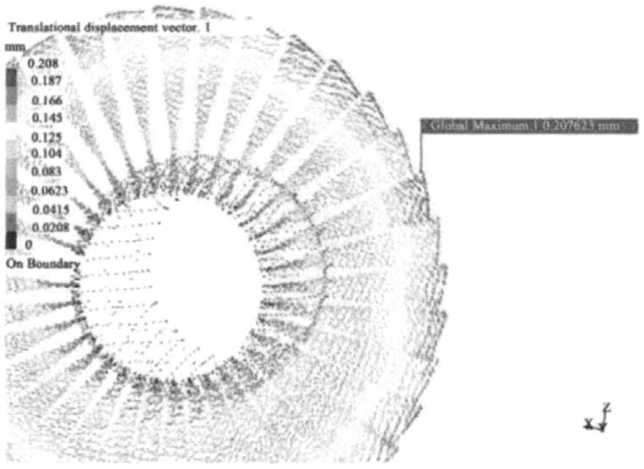

Figure 6. The stress deformation of fan at 800 N • m

From the foregoing, the force under different torque of the fan is basically the same, and the maximum stress position appears at the root of the blade, the maximum displacement deformation appears at the tip of the blade. Under the rated speed, the maximum radial displacement of the blade tip is 0.327 mm.,The greater the torque, the greater the stress, the stress position is basically the same.

According to the calculation results of 2.3.2 and 2.3.3, the maximum stress of fan appears at the root of the blade when working, and the maximum stress of fan is 404MPa at the rated speed. According to the Spey engine stress standard (EGD-3) [3] [4], the fan as a key component of the aero-engine,which must be checked for the static strength of the blade in order to prevent various failures. Spey engine stress standards are: under all normal operating conditions, the average circumferential stress of the turbine disk does not exceed 75% of $\sigma_{0.1}$,under centrifugal and thermal loads, the circumfer- ential stress at the inner diameter does not exceed 95% of $\sigma_{0.1}$ [5]. In this paper, the yield strength of the fan at the design speed is 750 MPa, while the maximum stress of the fan at the rated speed is 404 MPa and less than 95% $\sigma_{0.1}$. Therefore, in this calculation condition, the turbine disk satisfies EGD-3 Static strength design requirements.

3 Conclusions

In this paper, CATIA's own finite element software is used to analyze the stress of CFM56 engine fan blades under centrifugal load. From the analysis results, it can be seen that the maximum point of deformation occurs at the tip of the fan and the high stress area appears in the root area of the blade.Under rated power, the maximum Von Mises stress caused by rotary centrifugal force is 398 Mpa, the stress caused by rotating torque is 6.4MPa, and the maximum stress is superimposed on 404.4 MPa.The stress induced by torque is very low relative to the stress caused by centrifugal force. The stress of the fan under normal working condition is mainly caused by centrifugal load.

The fan is the key technology of civil high bypass ratio turbofan engine,In the future, the direction of the research includes the curved blade, the stirrup blade and the separated blade.Noise reduction, resistance to external damage and blade tolerance are also the key problems in the design of large fans [6].The key technology includes: unsteady viscous aerodynamic design technology; splitter blade design; swept fan blade design and manufacturing technology; blisk design and manufacturing technology; two flow control technology; brush seal design and manufacturing technology; the design of metal matrix composites.

The design of the future fan is mainly to continue improving the flow capacity, the blade tip cutting speed and the average level load.To improve the flow capacity can reduce the engine inlet diameter, so as to effectively reduce the weight of the engine, improve the thrust weight ratio. The method is to reduce the hub ratio (d) and increase the axial Maher number (Ma).At present, in the design of the fan, the wheel hub ratio is 0.34, and the Maher number is 0.631.Increasing the tangential speed and diffusion factor of blade tip can improve the stage pressure ratio, thereby reducing the required series under the given total pressure ratio and reducing the length and weight of the whole engine.

In recent years, with the continuous improvement of computer computing speed, finite element analysis has gained more and more attention in engineering design and analysis, and has become an effective way to solve complex engineering problems.Almost all of the design and manufacturing can not be separated from the finite element analysis and calculation.

References

1. LV Q. Strength analysis and life calculation of torque converter blades based on fluid-solid interaction. [D]. Nanjing. Nanjing University of Aeronautics and Astronautics, 2014.
2. Hou J D, Chen S X. Strength analysis of wide-chord fan blade from an aero engine[J].Aeronautical Computing Techniue, 2015,45 (2): 10-12.
3. Fu N. Strength analysis and life calculation of turbine disk and blade of an aero engine [D].Xi'an: Journal of Northwestern Polytechnical University, 2006.
4. Yu D Z , Wen D H, Zhang H R. Structural static analysis of gas turbine blade and disk of an aeroengine[J]. Structure & Environment Engineering. 2012, 39 (4):32-39.
5. Zhang M Q. Static strength calculation and analysis of the first stage turbine blade of turboprop six engine[J]. Journal of Civil Aviation University of China. 1983,1 (1): 29-39.
6. Wang C J, Song S G, Zong X. Vibration analysis of the blade disc coupled structure of compressor[J]. Journal of Aerospace Power. 2007,22 (7): 1065-1068.

A Novel Method on Min–Max Limit Protection for Aircraft Engine Control

Wenjun Shu[1], Bing Yu[1] and Hongwei Ke[1]

[1]*Nanjing University of Aeronautics & Astronautics, college of energy and power engineering, 29 Yudao Street, Nanjing, China*

Abstract. Min–Max selector structure is widely employed in current aircraft engine control logic. And the structure must provide desired thrust and prevent the engine from exceeding any safety or operational limits. In this paper, a new control scheme, that is Min–Max selector structure composed of Sliding Mode (SM) regulator and linear regulator, is presented. The main regulator is a linear regulator and all limit regulators are SM regulators. It could overcome the possibility of limit violation for the traditional Min-Max, and don' t need the augmented state references that is one drawback of SMC Min-Max(all regulators are SM regulator). The simulation results show that the proposed approach could effectively prevent limit violation and can improve Min–Max limit protection for aircraft engine control.

1 Introduction

During the tracking process of the aircraft engine control, it must be ensured that all the constrained output variables (such as temperature, surge margin, speed and pressure, etc.) should be within the allowable working range. Otherwise, it will affect the life of the engine or even cause appearance of the undesirable characteristics, which will cause the failure of the engine and a series of serious consequences[1, 2]. Therefore, how to play the best performance of aircraft engine under the limit protection becomes the key of its control research[3].

Traditional aircraft engine control system is usually controlled by the linear regulator using the Min-Max structure to achieve the constraint management[4-6]. In recent years, researches show that the traditional Min-Max linear control structure has certain conservatism in the constraint treatment, which hinders the dynamic response that may be faster under the constraint. Although Min-Max algorithm is known as a methodology to control output constrained systems, this does not imply that limited outputs remain within their limits during transient state, and therefore, limit violation may occur for some variables[7, 8]. Research conducted by Ritcher et al. [8, 9] indicated that many shortcomings of the standard min–max approach can be removed by replacing linear regulators with Sliding Mode Control (SMC). And in this paper, it is called SMC Min-Max. The constraint outputs are able to work closely in the transient process. And it improves the dynamic response performance of the system. However, this technique requires the specification of augmented state references, which includes target steady values for the actuators, since integral controls used. The steady map of the nonlinear engine must be used to pre-calculate such references. This drawback is aggravated when health parameter changes occur, since the steady map depends on these uncertain parameters. Ryan et al.[7] introduced the concept of using Conditionally Active (CA) limit regulators in the Min-Max architecture of typical engine control laws. The CA architecture in the Min-Max scheme only activates the limit regulators when the operating point is within a certain bound of the limit and approaching the limit at a faster rate than the prescribed one. Imani et al.[10] presented a strategy to design linear regulators of Min–Max selector control to improve transient limit protection by using Linear Matrix Inequalities (LMI) approach. It decreases the possibility of transient limit violation and can be considered as an effective approach to improve Min-Max limit protection in aircraft engine control. But the strategy did not show an advantage in response speed.

In this paper, a Min-Max structure controller composed of linear regulator and SM regulator is used for aircraft turbofan engine, whose architecture is shown as Fig. 1. The main regulator is a linear regulator while all the limit regulators are SM regulators. The proposed approach don't need state references and the SM limit regulators will improve the response speed without the risk of exceeding the physical or safety limits.

2 The improvement of Min-Max controller

The turbofan engine is frequently controlled by a feedback loop where fuel flow rate is the control input and fan speed is the sensed variable. In this paper, a two-spool turbofan engine model in the 90,000 lb. thrust class, linearized at an altitude of 25,000 ft. and Mach number 0.62 is considered. And it is obtained from the

https://doi.org/10.1051/matecconf/201816604002

Commercial Modular Aeropropulsion System Simulation (C-MAPSS)[10, 11]. Table 1 shows some output variables and their units. The linear model is shown as follows:

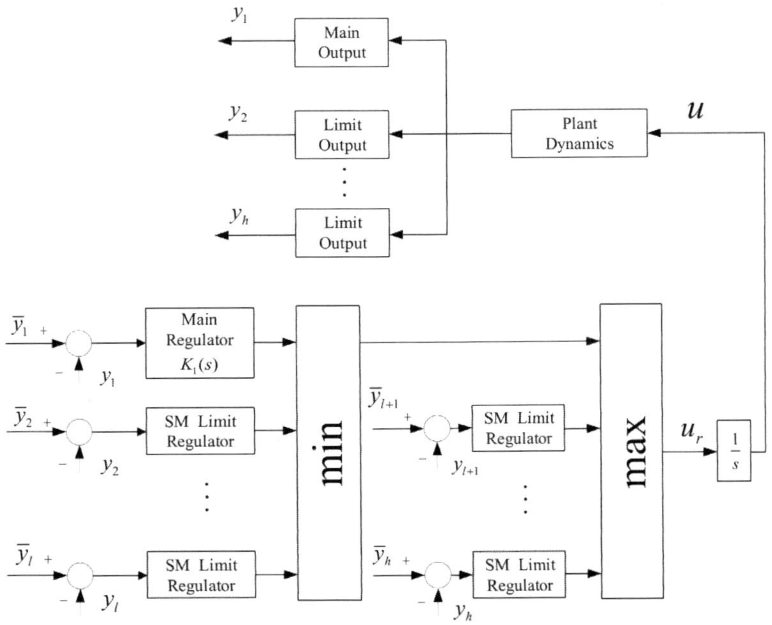

Figure 1. The new Min–Max multi-regulator system with integral control.

$$\dot{\boldsymbol{x}} = \boldsymbol{A}\boldsymbol{x} + \boldsymbol{B}u$$
$$\dot{u} = u_r \qquad (1)$$
$$y_i = \boldsymbol{G}_i\boldsymbol{x} + \theta_i u$$

Where $\boldsymbol{x} = [\Delta N_f \;\; \Delta N_c]^T$ is state variables vector with fan speed increment and core speed increment, $u = \Delta W_f$ is fuel flow increment as control input, and y_i, for i =2 , 3 , 4 , 5 are output increments. The model matrices come from references[10, 11].

Table 1. Some output variables of aircraft engine.

Variables	**Units**
Fan speed, N_f	rpm
Core speed, N_c	rpm
Fuel flow rate, W_f	pps
High-pressure compressor stall margin, $SmHPC$	%
High-pressure turbine outlet temperature, T_{48}	$^{\circ}R$
Engine pressure ratio, EPR	-
High-pressure compressor outlet pressure, P_{s30}	psia

The control objective is to follow the desired fan speed while maintaining constrained outputs within bounds at all times. The outputs and their incremental limits come from references[10, 11].

SMC Min-Max requires the specification of augmented state references, which includes target steady values for the actuators. However, that real-time implementations of the SM limit regulators do not require state references, due to the definition of the sliding function. Only the limit setting and a real-time output measurement are needed to calculate the sliding function[9]. To overcome this shortcoming, the main regulation is replaced by a linear regulator. Meanwhile, all limit regulations are SM regulators that will improve the response performance without exceeding the limits.

As shown in Fig.1, the linear regulator $K_1(s)$ is used as main regulator. Its parameter values come from the reference [10].

$$K_1(s) = \frac{0.74s+0.8}{s+40} \qquad (2)$$

In order to ensure that the limited output does not exceed its limit, SM limit regulators are used for all the limits. Due to space limitation, the method of SMC will not be repeated and just is given brief introduction[11]. The sliding function is defined by $s_i = y_i - \overline{y_i}(i = 2,3,4,5)$, and $\overline{y_i}$ is the physical or safety limit. The control law of SM limit regulator is given below in Eq. (3).

$$u_{ri} = -\frac{1}{\Theta_i}(\boldsymbol{G}_i(\boldsymbol{A}\boldsymbol{x} + \boldsymbol{B}u) + \eta_i sgn(s_i)) \qquad (3)$$

Where, η_i is switching gain, and it directly affects dynamic response; $sgn(\cdot)$ is saturation function. In this paper, the switch gain is set to $\eta_i = 15|\theta_i|$ (i = 2,3,4,5).

3 Simulation results and discussion

As the engine thrust cannot be sensed in a reliable way, fan speed variable is considered as thrust representative in this study. The performance is compared with the traditional Min–Max algorithm with controller and the LMI Min-Max controller. And all controllers are digital controllers with the sampling time of 0.015 seconds[6].

Fig.2 shows the system response in tracking the desired fan speed. Although settling time for both is three identical, rise time for the new controller is less than the LMI one and keeps the same as traditional one. Incremental fuel flow input has been shown in Fig. 3 for all controllers.

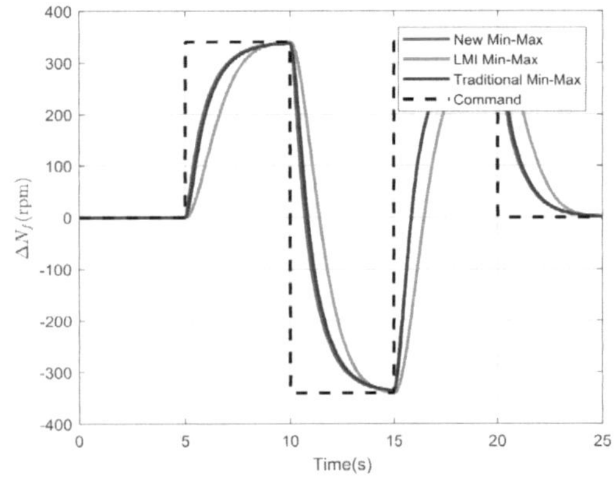

Figure 2. Fan speed response.

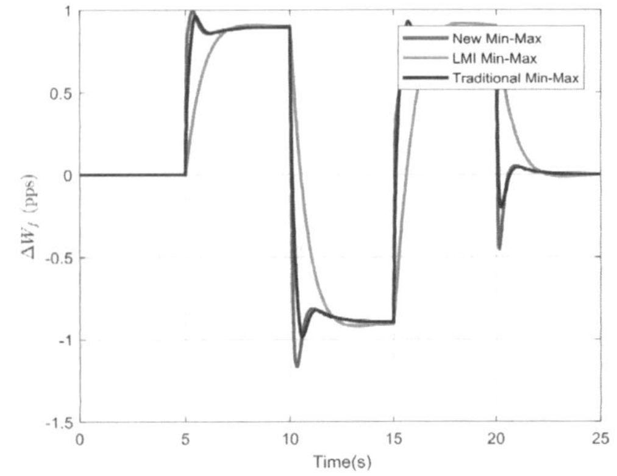

Figure 3. Fuel flow input.

Fig. 4 shows the output $\Delta \boldsymbol{SmHPC}$ and its detail with enlarged scale. Although rise time of new controller and traditional are same, it's obvious that the new proposed controller protects the constraint unlike the traditional. Fig. 5, Fig. 6 and Fig. 7 are the outputs of $\Delta \boldsymbol{T}_{48}$, $\Delta \boldsymbol{P}_{s30}$ and $\Delta \boldsymbol{EPR}$. All controllers satisfy the constrains.

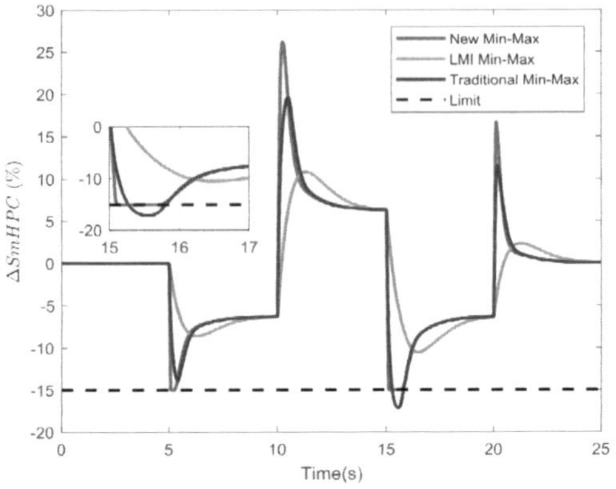

Figure 4. Lower limited output $\Delta \boldsymbol{SmHPC}$.

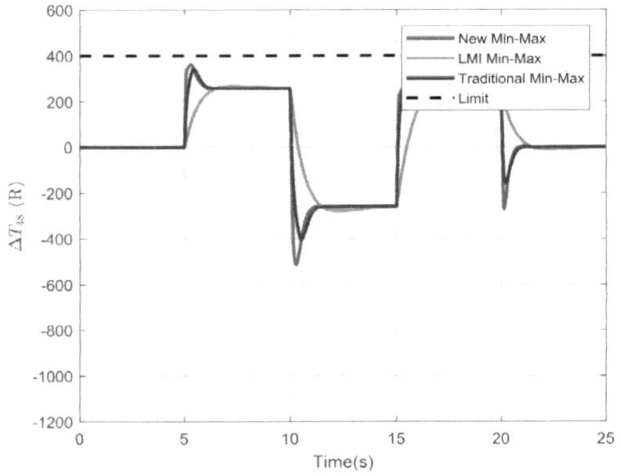

Figure 5. Upper limited output $\Delta \boldsymbol{T}_{48}$.

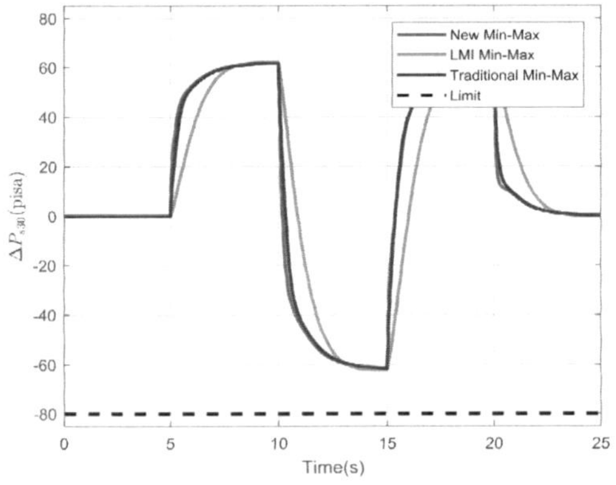

Figure 6. Lower limited output $\Delta \boldsymbol{P}_{s30}$.

https://doi.org/10.1051/matecconf/201816604002

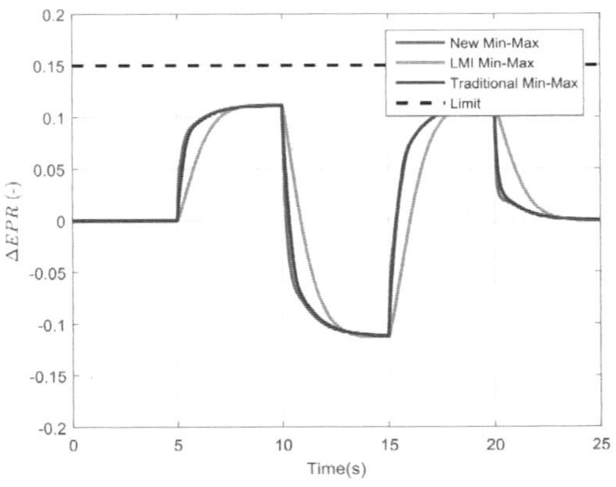

Figure 7. Upper limited output $\Delta \boldsymbol{EPR}$.

Because the SM limit regulators are used, the constraint outputs are rapidly approaching the limit value in the response without exceeding the limit value. It's proved that the new controller has the potential to reach the reference value faster. However, all of the above simulation results can not well show this advantages. Now, the linear main regulator will be redesigned that is shown in follow:

$$K_1(s) = \frac{0.74s+1.8}{s+40} \tag{6}$$

It adds a new comparison object that is SMC Min-Max means all the regulators are SMC structure, and others are same as the previous simulation.

Fig. 8 shows the system response in tracking the desired fan speed after the main regulator redesign. The settling time for both of new Min-Max and SMC Min-Max controller is almost identical, and both of them are faster than traditional and LMI Min-Max controller.

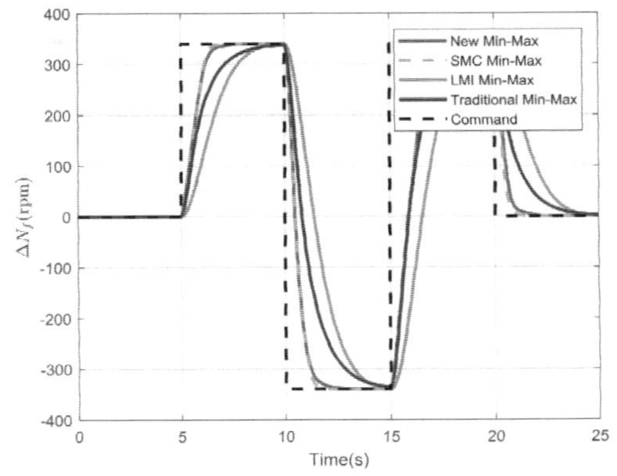

Figure 8. Fan speed response after the main regulator redesign.

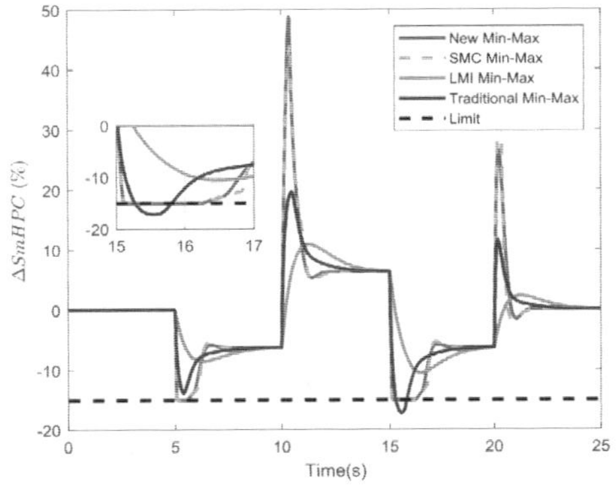

Figure 9. Lower limited output $\Delta \boldsymbol{SmHPC}$ after the main regulator redesign.

Due to the space limit, the constraint output only $\Delta \boldsymbol{SmHPC}$ is shown in Fig. 9. And it demonstrates that all the controllers except the traditional Min-Max controller prevent limit violation.

4 Conclusion

In this paper, a new control scheme, that is Min–Max selector structure composedof Sliding Mode (SM) regulator and linear regulator, is presented. The main regulator is a linear regulator and all limit regulators are SM regulators. It overcomes the possibility of limit violation for the traditional Min-Max, and don't need the augmented state references which is one drawback of SMC Min-Max. The simulation results show that the proposed approach could effectively prevents limit violation and improve Min–Max limit protection in aircraft engine control.

Acknowledgments

This work was supported in part by the Natural Science Foundation of China under Grant 51406083, in part by the Fundamental ResearchFunds for the Central Universities under Grant NJ20160037, and in part by the Natural Science Foundation of Jiangsu Province underGrant BK20140820.

References

1. Jaw, L.C. and J.D. Mattingly, *Aircraft engine controls : design, system analysis, and health monitoring*. American Institute of Aeronautics and Astronautics (2009)
2. Du, X., Y.Q. Guo, and X.L. Chen, Limit management of aircraft engine based on nonlinear model predictive control method. Journal of Aerospace Power. **30**(7): 1766-1771 (2015)
3. Csank, J.T. and R.D. May, A Sensitivity Study of Commercial Aircraft Engine Response for Emergency Situations: NASA/TM-2011-217004 (2011)

4. May, R., et al., A High-Fidelity Simulation of a Generic Commercial Aircraft Engine and Controller: AIAA-2010-6630 (2010)
5. Iii, H.A.S. and H. Brown, Control of jet engines. Control Engineering Practice. **7**(9): 1043-1059 (1999)
6. Csank, J., et al., Control Design for a Generic Commercial Aircraft Engine: AIAA-2010-6629 (2010)
7. May, R.D. and S. Garg, Reducing Conservatism in Aircraft Engine Response Using Conditionally Active Min-Max Limit Regulators: AIAA-2012-217814 (2012)
8. Richter, H., Multiple Sliding Modes with Override Logic: Limit Management in Aircraft Engine Controls. J GUID CONTROL DYNAM. **35**(4): 1132-1142 (2012)
9. Richter, H., *Advanced Control of Turbofan Engines*. Springer New York (2012)
10. Imani, A. and M. Montazeri-Gh, Improvement of Min-Max Limit Protection in Aircraft Engine Control: An LMI Approach. AEROSP SCI TECHNOL. **68** (2017)
11. Richter, H., A multi-regulator sliding mode control strategy for output-constrained systems. AUTOMATICA. **47**(10): 2251-2259 (2011)

Optimized Matching Lift Unit Transmission Ratio of Engine Driven Ducted Fan

Senlin Xiao , Hongxin Qiao , Guanxiong Xu , Guailin Luo , Chao Ma , Zhiyong Guo and Depeng Zhang

Kuang-chi Institute of Advanced Technology, No. 9 Gaoxin Middle 1st Road, 518100 Shenzhen, China

Abstract. As a kind of VTOL technology, ducted fan is not only used by many kinds of aircrafts, but also one of the trends of the future aircraft lift system, and attracts more and more attention. For an engine driven ducted fan lift unit, involving the engine and ducted fan matching problem, the form of transmission and transmission ratio are the key design parameters. In order to design and develop a ducted fan aircraft reasonably, a thrust test platform was set up to connect the engine with the ducted fan through the belt driving. The matching relationship between the engine and the transmission system was experimentally studied and the optimal transmission ratio was determined. The results showed that the optimal transmission ratio for the engine 1 is 2.2:1, and for the engine 2, the optimal transmission ratio should be 2.95:1 based on the current ducted and movable blade aerofoil design. At this time, the lift will exceed 130 kg•f, meeting the aircraft's original design requirements.

1 Introduction

The traditional fixed-wing aircraft, which requires high starting conditions and poor manoeuvrability, however ducted fan-powered aircraft has high-speed flying and hovering ability, as well as VTOL (Vertical Take-off and Landing) capability, less affected by the environment and good manoeuvrability, and has a wide range of applications, has become one of the important research hot spots [1-4]. Since the 1960s of the last century, ducted fan powered aircraft such as Airgeep, Doak 16, Cypher and AirMule have been released[1,5]. The latest news shows that Jetpack manufactured by Martin Jetpack in New Zealand is an aircraft with twin ducked fan as a lifting unit (see Figure 1)[6].

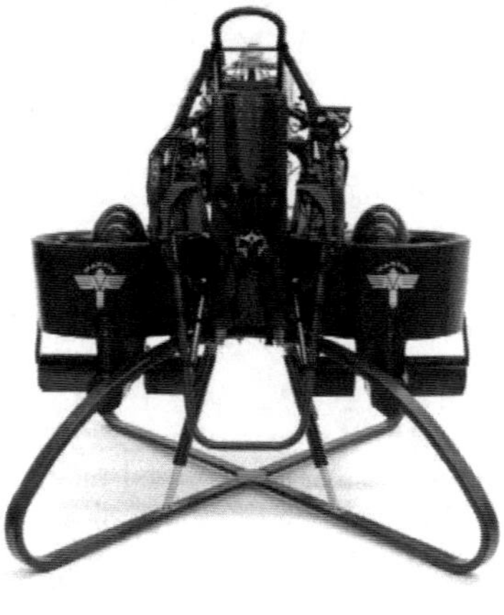

Figure 1. The latest version of Jetpack.

Early NASA studied the ducted fan aerodynamic characteristics and flow characteristics through the wind tunnel test. Ohanian III[3] investigated the potential of using Synthetic Jet Actuators (SJA) to control the flight stability of ducted fans. Will[4] experimentally studied the aerodynamic characteristics of various ducted lip shapes, including the leading edge radius of the ducted lip and the culvert wall thickness, and found that the smaller leading edge radius favours the stability under take-off and crosswind conditions, whereas the larger leading edge radius enhances the stability under static conditions. Further flow visualization and static pressure measurements revealed that the cause of the above differences was due to the flow separation inside the ducts lip surface. Pflimlin[5] designed an attitude control method based on a linearization of the dynamics model to control the balance stability of hovering of ducted fan micro UAVs in crosswind.

The lift unit formed by the engine-driven ducted fan relates to the match between the engine and the ducted fan, and the form of the transmission and the setting of the transmission ratio are directly related to the design of the aircraft and the performance of the aircraft. In this paper, a thrust test platform was set up which connected the engine with the ducted fan through the belt drive. The aerodynamic characteristics of the ducted fan in the hovering state were measured. Specifically, the relationship between the ducted fan suction power, the fan speed and the thrust generated by the ducted fan was measured. Then, a design idea of lift unit transmission system for engine-driven ducted fan was established and the transmission ratio was optimized, and evaluated the applicability of ducted fan system as the lift system of aircraft.

2 Thrust test platform

Thrust test bench, shown in Figure 2, length × width × height is 1900 mm × 1900 mm × 2334 mm, which including ducted fan, power systems, thrust gantry, transmission systems and sensors. Ducted fan is fixed on the mobile frame of thrust gantry. The power system is located under the ducted fan, fixed on the fixed frame of thrust gantry, connected to the fan shaft through the transmission system. The transmission system mainly includes two synchronous belt pulleys, one timing belt, two belt tension pulleys, and one transmission shaft. The changes of the transmission ratio are realized by changing synchronous pulleys with different numbers of teeth, and meanwhile the belt tensioning pulley is used to adjust the belt preload. The entire transmission system is shown in Figure 3. The mobile frame is connected with the heavy-duty linear guide on the fixed frame through the precision sliding block, so that the mobile frame can only move back and forth along the axial direction of the guide rail. Fixed frame is fixed to the ground, once the ducted fan is running, the thrust will be generated which causing the mobile frame to move along the guide rail, and the thrust will be measured by the thrust sensor.

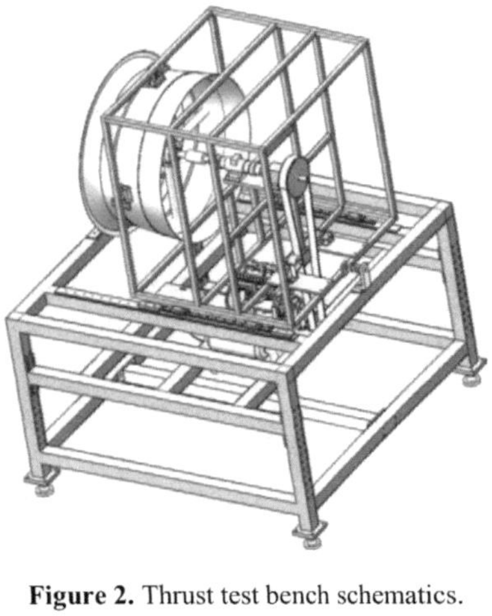

Figure 2. Thrust test bench schematics.

Figure 3. Transmission system diagram.

3 Test results and analyses

3.1. Engine performance

The performance of engine 1 and engine 2 is shown in Figure 1 and Figure 2. Obviously, the engine 2 can run at a higher speed range. The match between the engine and the ducted fan must consider the performance characteristics of different engines.

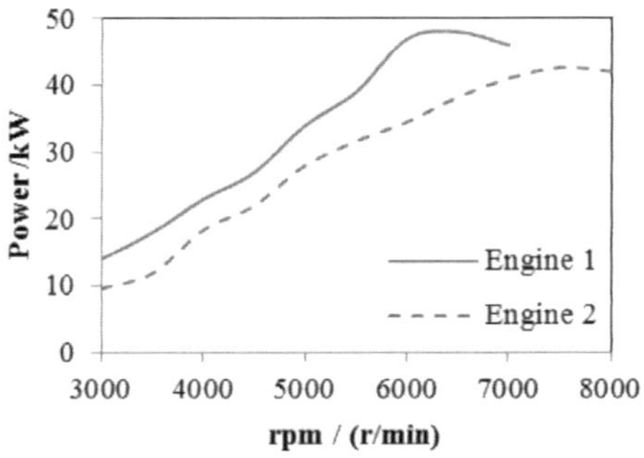

Figure 4. Power of engine 1 and engine 2.

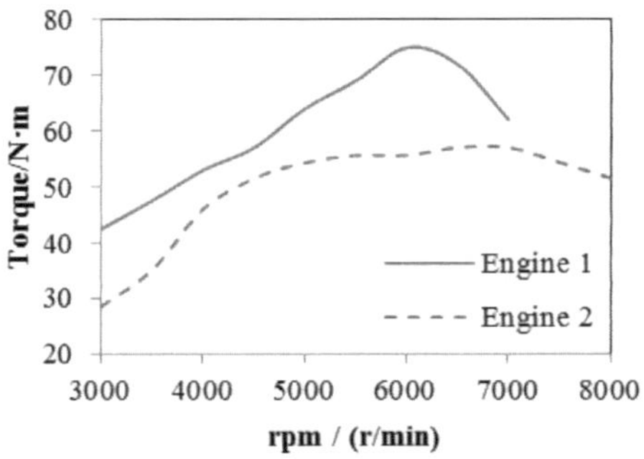

Figure 5. Torque of engine 1 and engine 2.

3.2. Matching analysis

In order to avoid the impacts of environmental conditions on the test results, ducted fan absorption power P_c was calculated according to equation 1, the measured data during the test included the fan speed n_f and torque M_e. Ducted fan thrust F_0 was calculated by Equation 2, where T was measured ambient temperature, T_0, p_0 was the international standard atmospheric temperature and pressure; T, p was the measured ambient air temperature and pressure; F was the measured thrust.

$$P_c = 2\pi\, n_f\, M_e / 60 \tag{1}$$

$$F_0 = (T / T_0)\,(p_0 / p)\, F \tag{2}$$

The design of the transmission relationship is based on the structural design. At the same time, the form and matching of the transmission system directly affect the aerodynamic load distribution of the ducted fan. Furthermore, the distribution of the aerodynamic load will in turn affect the structural design. The matching principle of the transmission system is that on one hand, the engine runs a suitable target operating point, and gives full play to the performance of the engine. On the other hand, the ducted fan runs at the best working point, and provides suitable lift for the entire lifting unit. And the thrust-to-weight ratio of the entire lift unit meets the design goals.

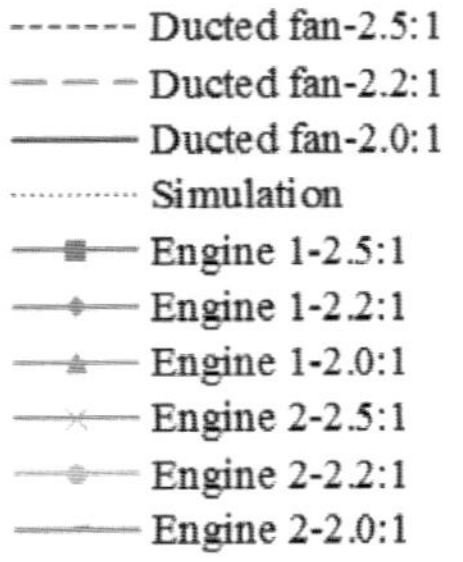

Figure 6. Illustration of each figure.

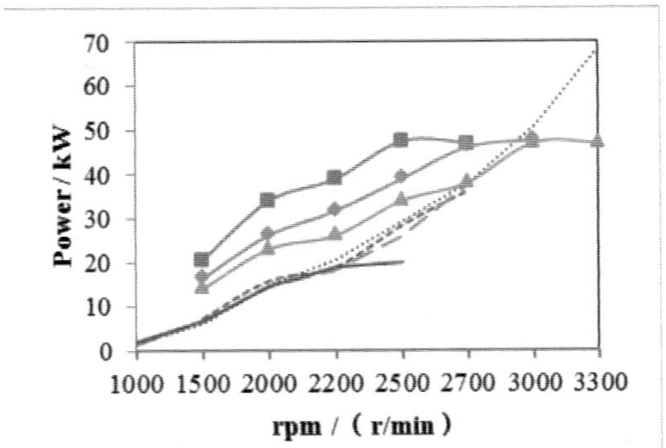

Figure 7. The relationship between power and speed when the engine 1 matches the ducted fan.

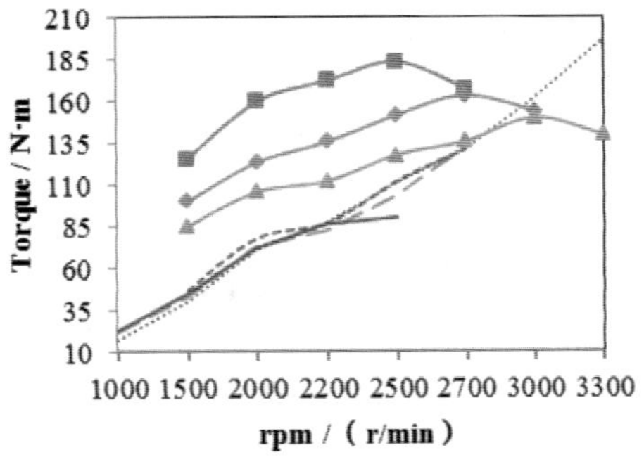

Figure 8. The relationship between torque and speed when the engine 1 matches the ducted fan.

Figure 7 and Figure 8 are the relationships between power and speed, torque and speed when engine 1 matches the ducted fan. The figure gives the engine speed characteristics when the gear ratios are 2.0:1, 2.2:1 and 2.5:1, respectively. For comparative analysis, Figure 6 shows the legend in the following figures. At the same time, the figure shows the aerodynamic characteristics calculated by CFD as a reference and characteristics tested by thrust test bench. The test verified the rationality and reliability of the simulation and confirmed that the simulation data can be used to design and optimize the ducted fan. Note that the test data is represented by ducted fan-2.5:1, ducted fan-2.2:1, and ducted fan-2.0:1 respectively in figures. In addition, the influence law of the propeller blade pitch angle on the thrust was studied experimentally. For current duct and blade aerofoil designs, the optimal blade pitch angle is 30°. In the next study, the pitch angle remained unchanged, so the ducted fan's thrust characteristics and the speed formed a fixed relationship. Obviously, it can be found that the matching of the engine 1 and the ducted fan is most reasonable when the transmission ratio is 2.2:1. At this ratio, the ducted fan can operate at higher speeds with sufficient power and torque. The significant advantage is that the performance of the engine can be effectively utilized while the ducted fan's thrust can be increased accordingly.

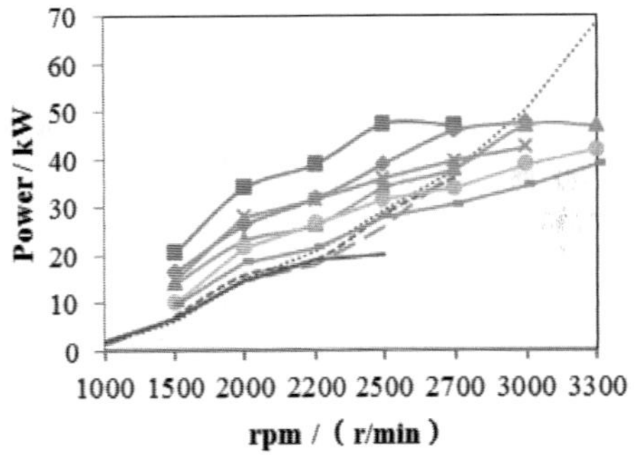

Figure 9. The relationship between power and speed when the engine 2 matches the ducted fan.

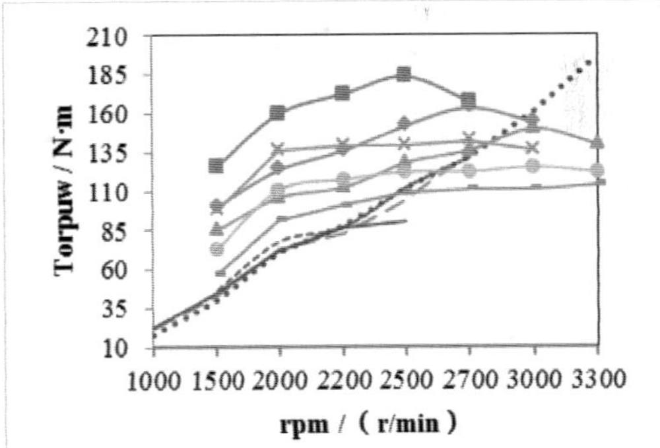

Figure 10. The relationship between torque and speed when the engine 2 matches the ducted fan.

Figure 9 and Figure 10 are the relationships between power and speed, torque and speed when engine 2 matches the ducted fan. The figure gives the speed characteristics of engine 2 when the gear ratios are 2.0:1, 2.2:1 and 2.5:1, respectively. When the transmission ratio was 2.5:1, the fan speed was 2500 r/min, output of engine 2 was 36 kW, and the measured duct fan absorption power was 28.477 kW. Therefore the power exceeded the demand of about 26.418%, and the torque exceeded the demand of about 15.970%. When the transmission ratio was 2.2:1, the fan speed was 2500 r/min, output of engine 2 was 31.6 kW, and the measured duct fan absorption power was 25.95 kW. Therefore the power exceeded the demand of about 21.773%, and the torque exceeded the demand of about 17.615%. However, the speed of ducted fan couldn't exceed 2700 r/min because of lack of the efficient torque for the output of gear. When the transmission ratio was 2.0:1, the transmission output of torque was less than the ducted fan torque. In other words, the transmission ratio is invalid for the current construction. Taking into account the transmission efficiency and friction loss of whole system, the output torque of transmission must greater than the torque of ducted fan.

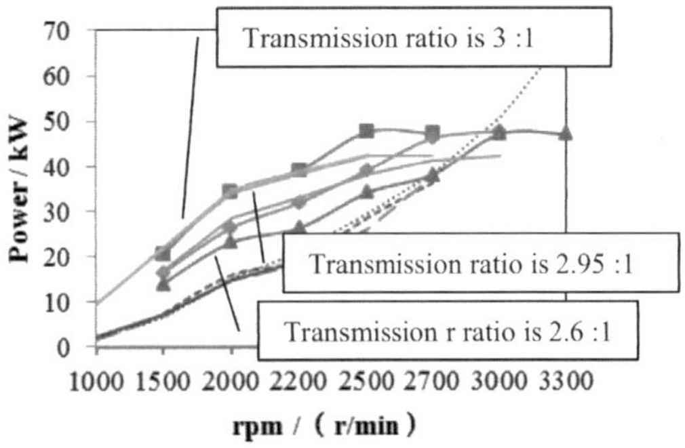

Figure 11. The relationship between power and speed when the engine 2 matches the ducted fan by increasing transmission ratio.

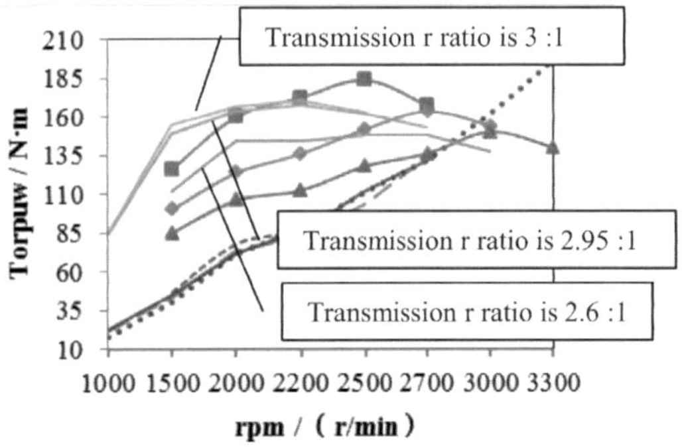

Figure 12. The relationship between torque and speed when the engine 2 matches the ducted fan by increasing transmission ratio.

Matching engine and ducted fan related to the performance of various subsystems, including the engine speed characteristics, engine load characteristics, ducted fan speed thrust characteristics, speed torque characteristics, different blade pitch angle characteristics, through constant optimization iteration of each parameter, it is necessary to design the working point and precise design of transmission ratio, which can significantly improve the overall performance of the aircraft. From the above analysis, we found that we can further increase the transmission ratio to match the engine and ducted fan. Figure 11 and 12 are the relationships between power and speed, torque and speed when engine 2 matches the ducted fan. Due to the maximum speed of engine 2 is 8000 r/min, referring to the power and torque characteristics of engine 1, it is known that the transmission ratio is 2.6:1, and the ducted fan speed can reach 2500 r/min, the lift can exceed 100 kg•f. If the transmission ratio is 3:1 or 2.95:1, the fan speed can up to 2600 r/min at least. And changing the gear ratio makes it possible to reach to 2700 r/min for ducted fan speed. At this point, the engine delivered 152.545 N•m with a measured duct fan torque of 131.4 N•m and the torque exceeded the demand of about 16.092%. Note that, when the ducted fan speed is 2700 r/min, the lift force will exceed 130 kg•f, which meeting the original design requirements of the aircraft. As the transmission ratio changes, the output characteristics of transmission system will be changed accordingly. Because of the engine speed characteristics, the speed couldn't exceed the 2700 r/min for engine 2 when the transmission ratio was 3.0:1, which was shown in Figure 11 and Figure 12. The only choice for transmission ratio is 2.95:1 if the lift force needed to exceed 130 kg•f.

4 Conclusions

In order to measure the thrust characteristics, thrust test bench was designed and developed, and the key parameters of the lift unit were determined through experimental research, which will provide guidance for further research and development of new aircraft. Based on the above research, there are following important conclusions:

(1) On one hand based on the maximum performance of the engine, on the other hand based on further increasing ducted fan speed to increase the lift, after comparing the three gear ratios, it is found that the optimal transmission ratio is 2.2:1 for engine 1.

(2) The test found that when the ducted fan speed is 2700 r/min, the resulting lift will exceed 130 kg•f, meeting the initial design requirements of the aircraft. Therefore, when matching engine 2 to a ducted fan, the best transmission ratio should be 2.95:1 and the pitch angle is 30 °.

References

1. Yuanwei Li, Boqi Xi, Guoxing Yi, Changhong Wang. Journal of Harbin Institute of Technology **42**, 5 (2010)
2. Kai Liu, Fuchen Ye. Advances in Aeronautical Science and Engineering **6**, 2 (2015)
3. Ohanian III. Ducted Fan Aerodynamics and Modeling, with Applications of Steady and Synthetic Jet Flow Control. (Virginia Polytechnic Institute and State University, United State, 2011)
4. Will E G. Effects of duct lip shaping and various control devices on the hover and forward flight performance of ducted fan UAVs. (Virginia Polytechnic Institute and State University, United State, 2005)
5. Pflimlin J M, Binetti P, Soueres P, Hamel T., Trouchet D. Modeling and attitude control analysis of a ducted-fan micro aerial vehicle. Control Engineering Practice **18**, 3(2010)
6. http://martinjetpack.com/

MATEC Web of Conferences **166**, 04004 (2018) https://doi.org/10.1051/matecconf/201816604004
ICMAA 2018

Adaptive Engine Torque Compensation with Driveline Model

Jinrak Park[1], Seibum Choi[1,a] and Jiwon Oh[2]

[1]*KAIST, Mechanical Engineering Department, 34141 Daejeon, Korea*
[2]*Hyundai motor group, Research & Development Division, 150, Hyundaiyeonguso-ro, Namyang-eup, Hwaseong-si, 18280 Gyeonggi-do, Korea*

Abstract. Engine net torque is the total torque generated by the engine side, and includes the fuel combustion torque, the friction torque, and additionally the starter motor torque in case of hybrid vehicles. The engine net torque is utilized to control powertrain items such as the engine itself, the transmission clutch, also the engine clutch, and it must be accurate for the precise powertrain control. However, this net torque can vary with the engine operating conditions like the engine wear, the changes of the atmospheric pressure and the friction torque. Thus, this paper proposes the adaptive engine net torque compensator using driveline model which can cope with the net torque change according to engine operating conditions. The adaptive compensator was applied on the parallel hybrid vehicle and investigated via MATLAB Simcape Driveline simulation.

1 Introduction

Recently, eco-friendly automobile technology is attracting attention. So, the automotive precision control technology is developing with electronic control system to satisfy the fuel economy and the performance of a vehicle. The accurate monitoring of operating states of a car is necessary for the precise control but sometimes it is impossible to utilize some sensors due to the lack of cost and space. Thus, the use of state observers using physical models is increasing. Typically, torque sensors are expensive and bulky, so they cannot be mounted on commercial vehicles. Thus, all torque states such as engine torque, engine clutch torque, transmission torque, motor torque, etc. should be estimated using other sensors.

Engine net torque is the torque measured at the crankshaft flywheel and is the sum of the fuel combustion torque, the friction torque and, additionally in case of hybrid vehicles, the starter motor torque. This net torque is the important state for the powertrain control such as the control of engine itself, engine clutch, transmission clutch, and for the vehicle dynamics control like active cruise control. However, the net torque cannot be measured because torque sensors cannot be utilized. So, it should be estimated. The fuel combustion torque of an engine is typically calculated through a map with the throttle opening and the engine speed. However, the map-based state estimator cannot correspond to the change of plant model and operating condition. The fuel combustion torque can vary with the personal structure tuning, the engine wear (aging), atmospheric pressure, etc. Also, the friction torque can change depending on the condition of oil and the engine wear. And, it should be also estimated. The engine net torque can be estimated by summing the fuel combustion torque and the friction torque. In case of hybrid vehicles, the starter motor torque is added to the engine net torque and it can be estimated somewhat precisely using the current sensor. However, various factors can affect the error of the engine net torque, so it is difficult to find out the cause of the error.

S. B. Choi, and J. K. Hedrick proposed the fuel combustion torque model which is accurate in both the steady and transient state [1]. Also, the engine net torque is compensated using the load torque which is assumed to be known, and driveline model. However, in commercial vehicles, it is difficult to know the exact load torque because the air drag force, the rolling resistance and the gradient resistance cannot be exactly estimated. Thus, another solution is needed to compensate the engine net torque instead of the method using the load torque.

J. Oh, J. Kim, and S. B. Choi suggested the adaptive estimator to estimate the quantity of the engine net torque compensation and the output shaft torque simultaneously using only the driveline model [2]. However, the convergent speed of estimated states could be slow because two unknown states are esimtated using only one erorr variable.

This paper proposes the adaptive engine net torque compensator using the torque compliance model of the output shaft and the driveline model to compensate the slowly varying torque error depending on the engine model change. The compensator in this paper estimates only one unknown state, the multiplication factor on the engine net torque, using one error variable. The torque compliance model of a shaft is widely used to estimate the shaft torque but it needs rotational angle and speed information of a shaft which is not suitable for a real

[a] Corresponding author: sbchoi@kaist.ac.kr

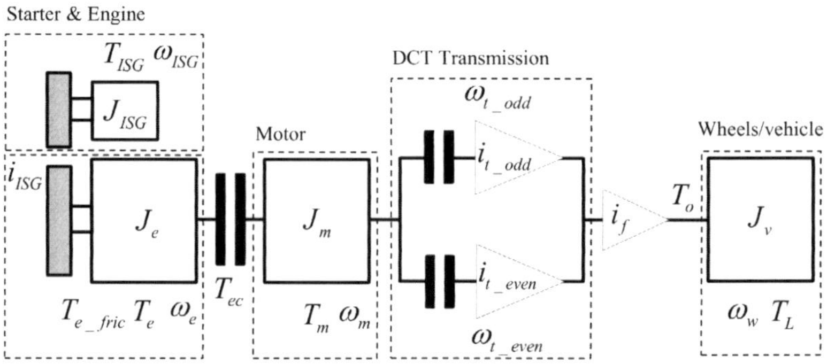

Figure 1. Driveline model of a parallel hybrid vehicle (*J*: inertia, *T*: torque, *ω*: angular velocity, *i*: gear ratio)

vehicle [3]. In case of using only speed information for commercial vehicles, the shaft torque is expressed as the differential form. Then, the drift issue can occur in estimating the shaft torque by integration. Thus, this paper also proposes the method to compensate the engine net torque while resolving drift issue on the shaft torque.

Also, the adaptive compensator is designed to use only sensors which can be utilized in a commercial vehicle. The adaptive algorithm is applied on a parallel hybrid vehicle in this paper but the algorithm can be modified easily for other types of a vehicle. The algorithm is investigated through MATLAB Simscape Driveline simulation. The organization of the paper is as follows.

Section 2 explains the detail algorithm of the compensator. Section 3 shows the simulation result. And, section 4 deals with the conclusion.

2 Detail algorithm of adaptive engine net torque compensator

2.1 Driveline model

Fig. 1 shows the lumped driveline model of a parallel hybrid vehicle and describes the torque flow. The mathematical model of a parallel hybrid vehicle which is utilized in this paper is as follows [4].

$$J_e\dot{\omega}_e = T_{e\_net} - T_{ec} \tag{1}$$

$$J_m\dot{\omega}_m = T_{ec} + T_m - \frac{T_o}{i_t} \tag{2}$$

$$\dot{T}_o = k_o\left(\frac{\omega_t}{i_t} - i_f\omega_w\right) \tag{3}$$

T_{ec}, T_o, k_o, i_t, and i_f mean the engine clutch torque, the output shaft torque, the equivalent torsional stiffness of the output shaft, the transmission gear ratio, and the final gear ratio.

The engine net torque is the sum of the fuel combustion torque, the starter motor torque, and the engine friction torque like below.

$$T_{e\_net} = T_e + T_{ISG} - T_{e\_fric} \tag{4}$$

T_{ISG}, and T_{e_fric} mean the starter motor torque, and the engine friction torque.

2.2 Adaptive engine net torque compensation algorithm

Using the mathematical driveline model of the parallel hybrid vehicle, the following equation can be reached by altering and summing (1) and (2).

$$T_{e\_net} = \frac{T_o}{i_t} - T_m + J_e\dot{\omega}_e + J_e\dot{\omega}_m \tag{5}$$

T_{e_net} indicates the actual engine net torque generated by the engine side.

Since the actual vehicle uses only the angular velocity sensor, the output shaft torque is expressed as a differential form as shown in equation (3), and drift occurs when the output shaft torque is obtained by integration. Thus, the torque variation formula in some specific period is utilized for the engine net torque compensation algorithm to escape from the drift issue like below.

$$\Delta T_{e\_net} = \frac{\Delta T_o}{i_t} - \Delta(T_m + J_e\dot{\omega}_e + J_e\dot{\omega}_m) \tag{6}$$

Assuming that the multiplied uncertainty exist on the nominal engine net torque, the engine net torque change can be expressed like below.

$$\hat{T}_{e\_net} = \hat{C}T'_{e\_net} \tag{7}$$

$$\Delta\hat{T}_{e\_net} = \hat{C}\Delta T'_{e\_net} \tag{8}$$

T'_{e_net}, and C mean the nominal engine net torque and the torque constant, and Δ indicates the amount of variation.

Then, the adaptive error is defined like below.

$$\begin{aligned}\varepsilon &= \Delta T_{e\_net} - \Delta\hat{T}_{e\_net} = (C - \hat{C})\Delta T'_{e\_net} \\ &= \Delta\left(\frac{T_o}{i_t} - T_m + J_e\dot{\omega}_e + J_e\dot{\omega}_m\right) - \hat{C}\Delta T'_{e\_net}\end{aligned} \tag{9}$$

The adaptive law of the torque constant can be defined using the simple gradient adaptive method like below equation [5].

$$\dot{\hat{C}} = \gamma\varepsilon\Delta T'_{e\_net}\ (\gamma > 0) \tag{10}$$

Defining the torque constant error like below,

$$\tilde{C} = C - \hat{C} \tag{11}$$

The following equation is satisfied using (10) and (11).

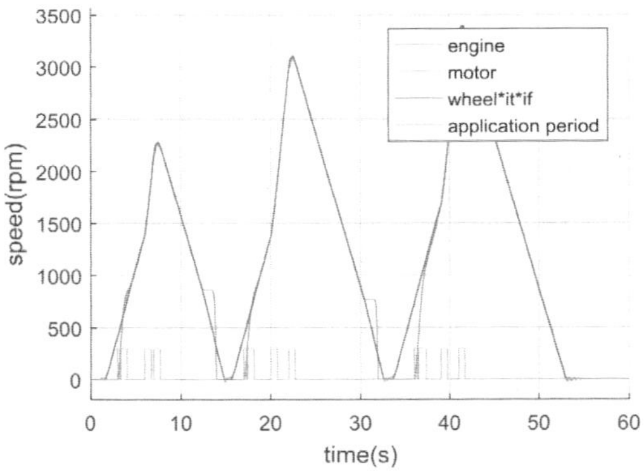

Figure 2. Rotational speed during simulation

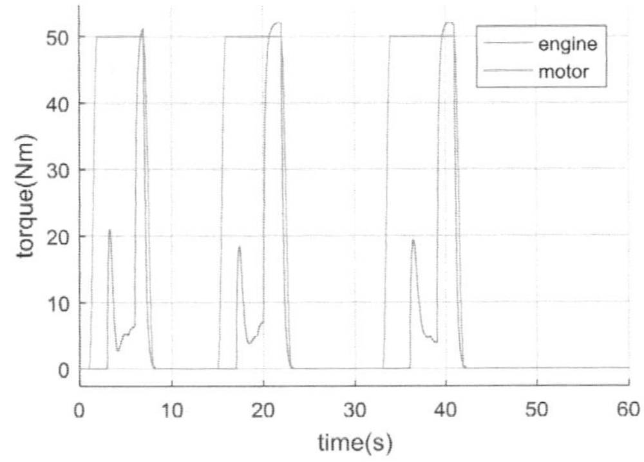

Figure 3. Torque during simulation

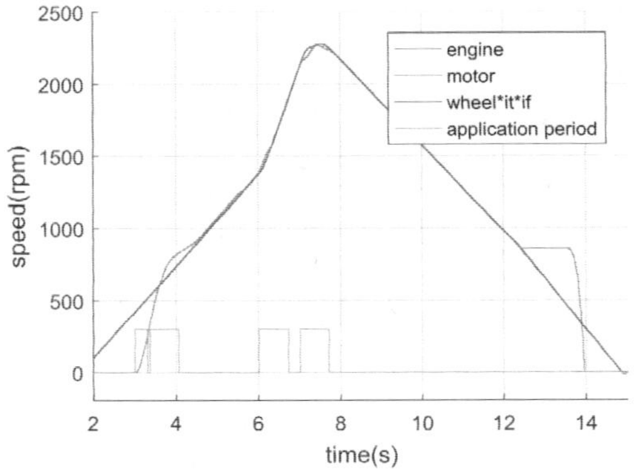

Figure 4. Enlarged graph of rotational speed during simulation

$$\dot{\tilde{C}} = -\gamma\tilde{C}\Delta T'_{e\_net}\ (\gamma > 0) \tag{12}$$

Then, the stability of the error dynamics can be proven using the simple Lyapunov candidate like below.

$$V(\tilde{C}) = \frac{\tilde{C}^2}{2\gamma} \tag{13}$$

$$\dot{V} = -\tilde{C}^2\Delta T'^{\,2}_{e\_net} \leq 0 \tag{14}$$

Since the variation of the nominal engine net torque satisfies the PE (persistence of excitation) condition, the adapted torque constant converges to the actual torque constant.

3 Simulation result

3.1 Simulation condition

The simulation in this paper was conducted using the Simscape Driveline tool in MATLAB Simulink. Fig. 2 shows the rotational engine speed of engine, motor, wheel speed which is multiplied the transmission and final gear ratio and, the application period of the adaptive algorithm. As mentioned in the section 2, the torque variation formula was utilized to address the drift problem of the output shaft torque in some specific period. Fig. 3 shows the generated torque by the engine side (engine net torque) and the motor. Also, Fig. 4 is the enlarged plot for some seconds of Fig. 2. For this simulation, it was assumed that the engine net torque is known but has some multiplied uncertainty and, the motor speed is the same with the transmission input shaft speed. Also, during the simulation period, the transmission clutch was engaged all the time and gear shift does not happened. Also, the engine and motor speed was synchronized three times.

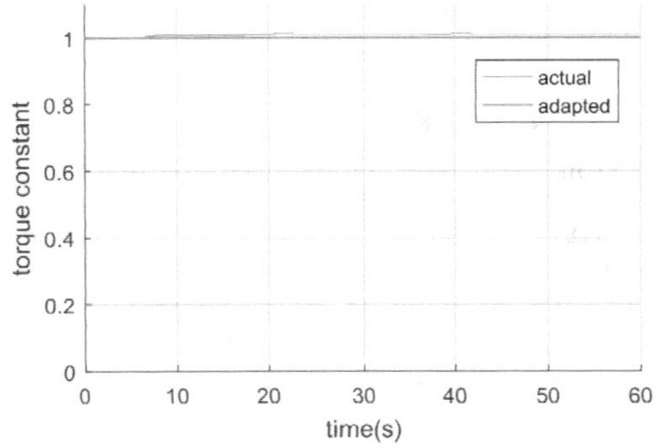

Figure 5. Result of torque compensator when the nominal torque constant is 1

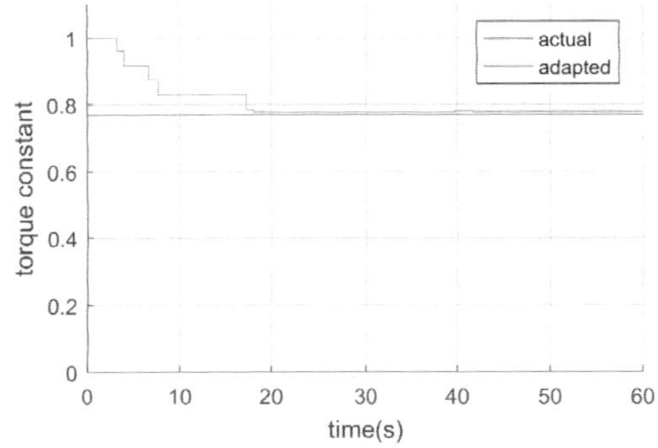

Figure 6. Result of torque compensator when the nominal torque constant is 0.77

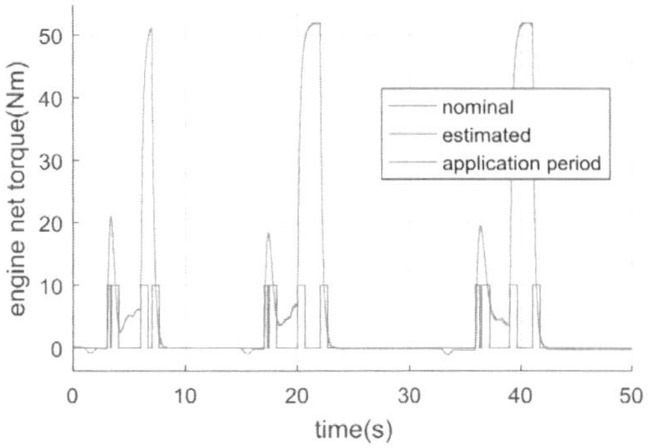

Figure 7. Comparison of engine net torque when torque constant is 1

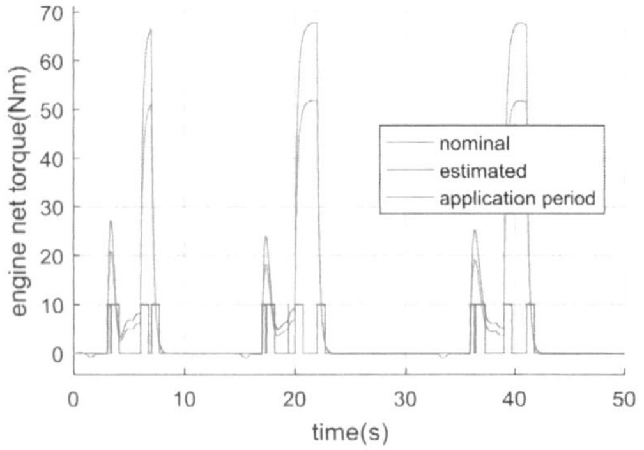

Figure 8. Comparison of engine net torque when torque constant is 0.77

3.2 Result

Fig. 5 shows the result of the torque compensator when the nominal torque constant is 1. Also, Fig. 6 shows the result when the nominal torque constant is 0.77. For the performance of the adaptive compensator, the adapted torque constant well converged to the actual torque constant but a little error, around 0.01, existed between the actual and adapted torque constant. It is because of the error on the output shaft torque. The torque compliance model of the output shaft is expressed in the differential form so the phase lag due to the integration existed. Also, the damping term was ignored in the compliance model.

Additionally, Fig. 7 shows the nominal and estimated engine net torque during the simulation when the nominal torque constant is 1. And, Fig. 8 shows the same thing when the nominal torque constant is 0.77. As you can see in Fig. 7 and Fig. 8, the adaptive torque compensator compares the difference of the nominal and estimated engine net torque, respectively, for some specific periods, and the torque constant is adapted using that difference error.

4 Conclusions

This paper proposed the adaptive engine net torque compensator according to the multiplied uncertainty on the net torque. This is to compensate the net torque error which is caused by the slow variation of the engine model. The engine model can change when the engine ages or wears, and can also be changed by an external environment, such as atmospheric pressure. The adaptive law of the torque constant for the engine net torque compensator was designed based on the variation of the engine net torque. The adaptive laws for the torque compensator was investigated through the simulation using the Simscape Driveline tool in MATLAB. The error between the actual and adapted torque constant occurred around in amount of 0.01. It is because of the error on the estimated output shaft torque. The improvement of this problem is left as a future work.

References

1. S. B. Choi, and J. K. Hedrick. "Robust throttle control of automotive engines: Theory and experiment." *Journal of dynamic systems, measurement, and control*, **118.1**, 92-98, (1996)
2. J. Oh, J. Kim, and S. B. Choi. "Design of estimators for the output shaft torque of automated manual transmission systems. " In *Industrial Electronics and Applications (ICIEA), **2013 8th***, 1370-1375, (2013, June).
3. J. Oh, and S. B. Choi. "Real-time estimation of transmitted torque on each clutch for ground vehicles with dual clutch transmission." *IEEE/ASME Transactions on Mechatronics*, ***20*(1)**, 24-36, (2015)
4. J. Oh, S. B. Choi, Y. J. Chang, and J. S. Eo. "Engine clutch torque estimation for parallel-type hybrid electric vehicles." *International Journal of Automotive Technology*, ***18*(1)**, 125-135, (2017).
5. P. A. Ioannou, & J. Sun. "Robust adaptive control." *Upper Saddle River, NJ: PTR Prentice-Hall*, **1**, (1996)

Acknowledgement

This work was supported in parts by the National Research Foundation of Korea grant funded by the Korea government (MSIP) (No.2017R1A2B4004116), the Ministry of Science and ICT, Korea, under the Information Technology Research Center support program (IITP-2017-2012-0-00628) supervised by the Institute for Information and communications Technology Promotion, and the BK21 Plus Program.

AUTHOR INDEX